How to build safer houses with confined masonry

Praise for this book

'This book is a great example of something good emerging from the tragedy of the 2010 Haiti earthquake. The existing reinforced concrete and masonry construction was essentially destroyed. So now, the safe alternative – confined masonry is explained in a way that masons can engage with. In a step-by-step detailed approach readers are instructed in how to build a house in confined masonry.

With its focus upon the practical skills and orientation of its readership almost all the content of the book is conveyed, not through text, but through attractive and well-annotated drawings. Even when communicating technical principles, a simple analogy gets the messages across.

This book warrants wide international dissemination to educate masons and others in the safest way to build houses using the most commonly available construction materials, reinforced concrete and masonry.'

Andrew Charleson, Associate Professor
in Building Structures, Victoria University of Wellington

'This unique guide illustrates construction of low-rise confined masonry buildings in a simple and user-friendly manner, and is expected to be an invaluable resource for house owners and builders of confined masonry houses in earthquake prone regions of the world.'

Dr. Svetlana Brzev, Chair, Confined Masonry Network,
Earthquake Engineering Research Institute

How to build safer houses with confined masonry

A guide for masons

Nadia Carlevaro, Guillaume Roux-Fouillet
and Tom Schacher

PRACTICAL ACTION
Publishing

Practical Action Publishing Ltd
The Schumacher Centre, Bourton on Dunsmore, Rugby, Warwickshire,
CV23 9QZ, UK
www.practicalactionpublishing.org

First published by Swiss Agency for Development and Cooperation and
Earthquake Engineering Research Institute 2015
This edition published by Practical Action Publishing 2018

A catalogue record for this book is available from the British Library.
A catalogue record for this book has been requested from the Library of
Congress.

ISBN 978-185339-989-3 paperback
ISBN 978-185339-988-6 hardback
ISBN 978-178044-988-3 library pdf
ISBN 978-178044-989-0 epub

Citation: Carlevaro, N. Roux-Fouillet, G and Schacher, T., (2018) *How to build
safer houses with confined masonry: A guide for masons*, Rugby, UK: Practical
Action Publishing <http://dx.doi.org/10.3362/9781780449883>

Since 1974, Practical Action Publishing has published and disseminated books
and information in support of international development work throughout the
world. Practical Action Publishing is a trading name of Practical Action Publishing Ltd (Company Reg. No. 1159018), the wholly owned publishing company of
Practical Action. Practical Action Publishing trades only in support of its
parent charity objectives and any profits are covenanted back to Practical
Action (Charity Reg. No. 247257, Group VAT Registration No. 880 9924 76).

The views and opinions in this publication are those of the author and do not
represent those of Practical Action Publishing Ltd or its parent charity
Practical Action. Reasonable efforts have been made to publish reliable data
and information, but the authors and publisher cannot assume responsibility
for the validity of all materials or for the consequences of their use.

Cover photo: A mason at work
Cover design: RCO.design
Credit: www.123RF.com Photographer: Patricia Hofmeester - The Netherlands
Typeset by vPrompt eServices, India
Printed in the United Kingdom

http://dx.doi.org/10.3362/9781780449883.000

Contents

PREFACE

This guide was originally developed by the Competence Center for Reconstruction of the Swiss Agency for Development and Cooperation (SDC) after the devastating January 2010 Haiti earthquake.

It was developed as a resource for the mason training programme for developing confined masonry construction skills. This training was launched as a response to the urgent need to establish an earthquake-resistant construction practice in Haiti. Its main purpose was to improve workmanship in areas where housing re-construction occurred without technical input.

This guide is regularly used at construction sites and as a resource material for mason training programmes. It offers simple but essential advice on building safer houses using the confined masonry construction technique.

This version of the Guide was adapted by SDC together with members of the Confined Masonry Network of the Earthquake Engineering Research Institute (EERI) for use in various countries and regions of the world.

It is hoped that this resource, originally developed in Haiti, will be useful in other countries facing similar challenges. It is intended for use by local governmental and non-governmental organizations, international humanitarian and development agencies, and most importantly skilled and unskilled masons around the world.

ACKNOWLEDGMENTS

All illustrations are by the authors and by other architects of the Competence Centre for Reconstruction of the Swiss Agency for Development and Cooperation (SDC) in Haiti, Martin Siegrist and Dorothée Hasnas.

We would like to thank those who gave their time and expertise to review this book: Dr Svetlana Brzev and Eng Tim Hart of the Confined Masonry Network (EERI); Marjorie Greene and Maggie Ortiz of EERI; Dr Andrew Charleson of the World Housing Encyclopedia (EERI) and Earthquake Hazard Centre.

ABOUT THE AUTHORS

Nadia Carlevaro and Guillaume Roux-Fouillet are architects and the founders of mobilstudio, with a decade of humanitarian experience in designing and training on earthquake and cyclone-resilient buildings in Myanmar, Haiti, the Philippines, Nepal and Ecuador. Both work regularly as construction and planning experts for the Swiss Agency for Development and Cooperation (SDC), the International Federation of the Red Cross and Red Crescent (IFRC) and the United Nations Refugee Agency (UNHCR).

Tom Schacher is an architect working regularly with the Swiss Agency for Development and Cooperation and has previously developed manuals and training materials for construction workers on locally appropriate earthquake-resistant construction techniques.

INTRODUCTION

How to Build Safer Houses with Confined Masonry is intended for the training of masons in the technique of confined masonry. It can be used as a guide on construction sites or as a training resource. It is presented in a simple manner and explains in a step-by-step sequence how to build a one or two-storey confined masonry house.

The guide was developed for masons working in countries with very limited financial and technical resources. The recommendations are intended to be conservative (on the safe side) and to ensure the safety of the occupants.

This guide needs to be adapted according to the type and quality of locally available materials and local capacities. The technical recommendations contained in the guide should be in compliance with local construction codes and other regulations (where available).

Illustrations included in the guide may be adapted to suit the local culture and perceptions and to ensure good acceptance. The text may be translated into a local language which the masons are able to read and understand.

While the authors have tried to be as accurate as possible, they cannot be held responsible for construction that might be based on the material presented in this guide. The authors and their organizations disclaim any and all responsibility for the accuracy of any of the material included in the guide.

http://dx.doi.org/10.3362/9781780449883.001

1. THE MASON'S WORLD

Masonry tools 1

   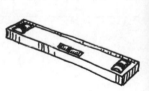

guide book   tape measure   straight edge   level

pencil   plumb line   string   nail   chalk line

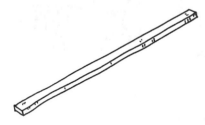   

aluminium screed   machete   screen (05, 03)

trowel   float   hammer   chisel   club hammer

Masonry tools 2

bucket

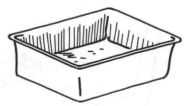
mixing box

cone for
slump test

big brush

transparent water
hose 10–20 m

pickaxe

shovel

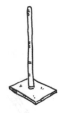
rammer

grinder

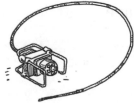
needle vibrator

concrete mixer

wheelbarrow

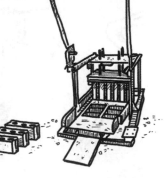
vibrating block/brick press

Formwork tools

 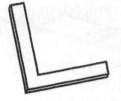 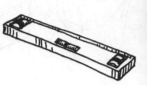

guide book   tape measure   straight edge   level

pencil   plumb line   string   nail   hammer   chisel

 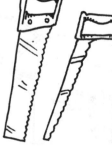 

crowbar   axe   saw   plane

Steel reinforcement tools

 guide book

 tape measure

 straight edge

 level

 pencil

 chalk

 plumb line

 string · nail

 wire twister
or pincer

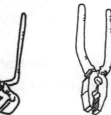 pliers

 tin snips

 hammer

 chisel

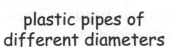 plastic pipes of
different diameters

 hacksaw

 rebar
bender

 chain bolt
cutter

Quality of materials

The quality of materials is essential to ensure
safe construction

Water: clean and
not salty

Blocks and bricks: (ch. 9)
minimal size and strength

Sand: river sand,
washed and dry

Cement: portland
cement, new and dry bags

Gravel: crushed or round,
from hard rock and clean, well-
graded, max size 18–20 mm

Steel bars: standard size,
ribbed steel, grade 60 new
and not corroded

Storage of building materials on site

Store cement bags away from the sun
and protected from humidity.
Do not place on the ground.

Store wood and steel bars in a dry environment.
Do not place on the ground.

Construction site protection

Do not forget that health and security concerns everybody, starting with yourself

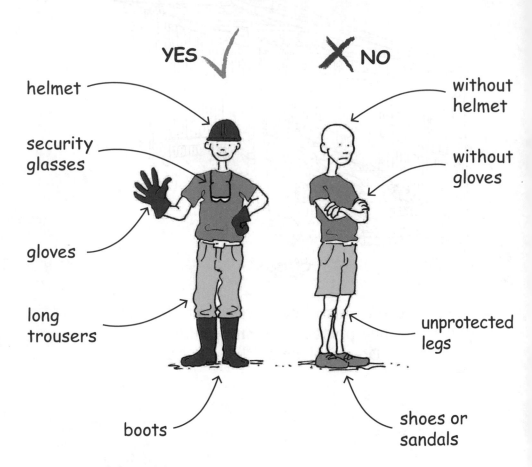

YES ✓ ✗ NO

helmet

security glasses

gloves

long trousers

boots

without helmet

without gloves

unprotected legs

shoes or sandals

If people are injured on a construction site, wash the wound with clean water and soap and go to a doctor.

http://dx.doi.org/10.3362/9781780449883.002

2. CONFINED MASONRY FOR TWO-STOREY HOUSES

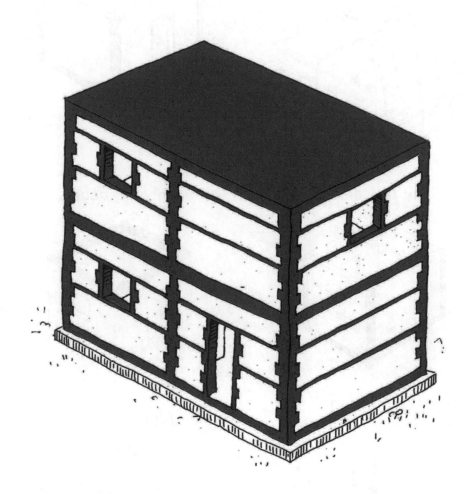

Confining elements (ties)

Confining the walls is like holding a pile of books together with a string: they can still move but they will not fall apart.

Horizontal ties (tie-beam) and vertical ties (tie-column).

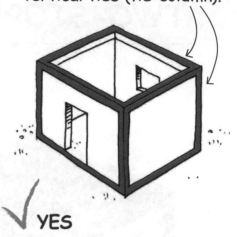

✓ YES

only tie-columns

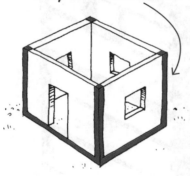

only tie-beams

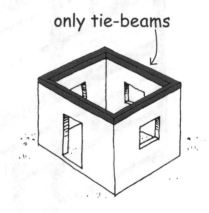

✗ NO

✗ NO

A strong house

All walls and openings should be confined to ensure stability during an earthquake.

Confining elements: (chapters 6–8) tie-column and tie-beams (plinth beam and ring beam)

Anchoring bands and opening reinforcement: (chapter 11) seismic bands (lintel and sill bands) and vertical reinforcement

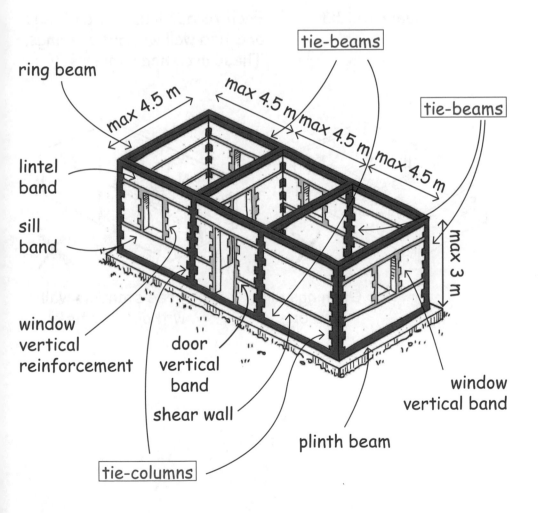

13

Shape of the house

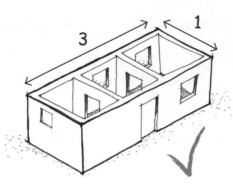

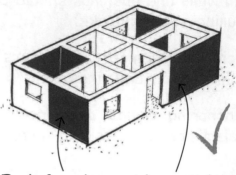

Maximum ratio 1:3.

Each facade must have at least one tied wall without openings. These are shear walls.

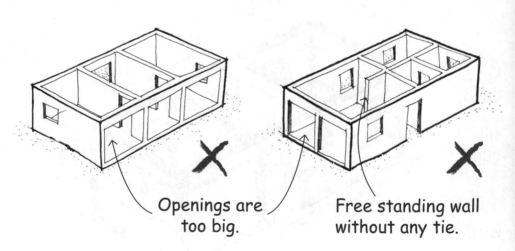

Openings are too big.

Free standing wall without any tie.

Shear walls

Shear walls are walls without windows or with a small window outside of the diagonals of the wall

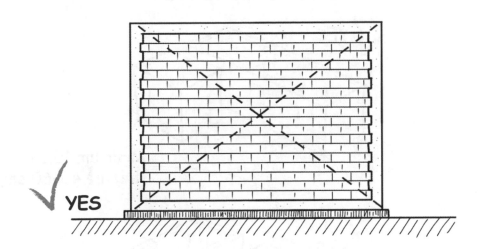

YES

Full shear wall

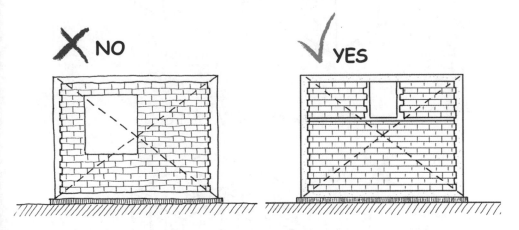

NO

YES

Opening is too big, crossing the diagonals: not a shear wall.

Opening is small and outside the diagonals: it is a shear wall.

Seismic gap

Avoid complex shapes by creating seismic gaps.

Simple shape: BETTER

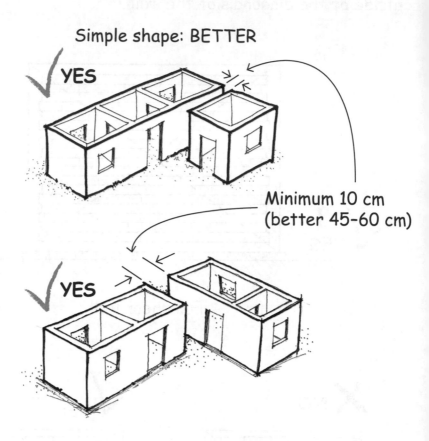

✓ YES

Minimum 10 cm
(better 45–60 cm)

✓ YES

Complex shape: WORSE

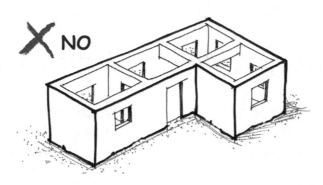

✗ NO

Vertical continuity of walls

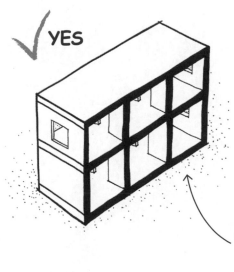

YES

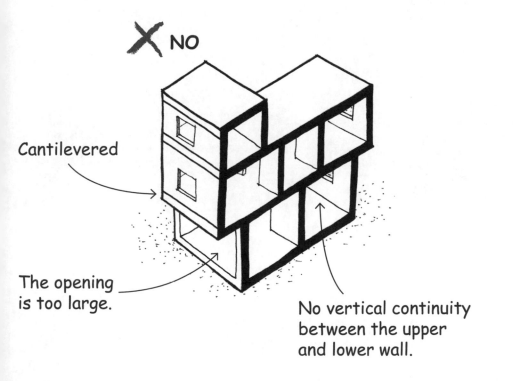

Walls must be placed continuously, one on top of the other from ground to the roof.

NO

Cantilevered

The opening is too large.

No vertical continuity between the upper and lower wall.

3. FINDING AN ADEQUATE LOCATION

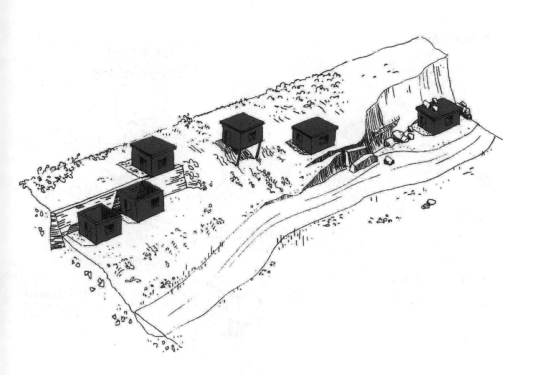

Site selection: where to build

YES

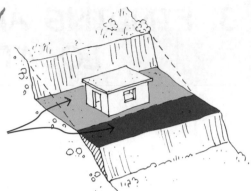

Keep enough distance on each side of the house.

Don't build on embankments.

NO

Don't build on fresh embankments.

NO

Don't build on stilts.

NO

Don't build too close to a cliff.

NO

Don't build at the foot of a cliff.

NO

Flood-related hazards

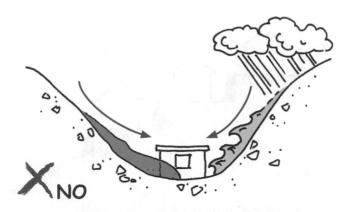

X NO

Don't build at the bottom of a canyon.

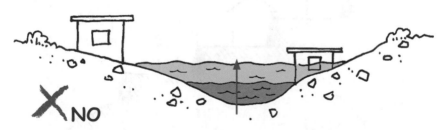

X NO

Don't build near a river.

X NO

Don't build near the ocean
(due to tsunami risk).

Building on a slope

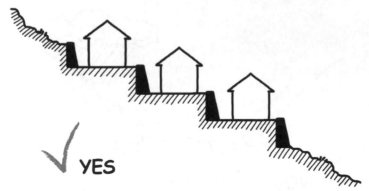

✓ YES

Build between retaining walls.

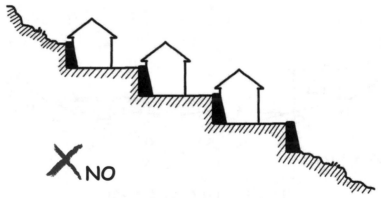

✗ NO

Don't build against a retaining wall.

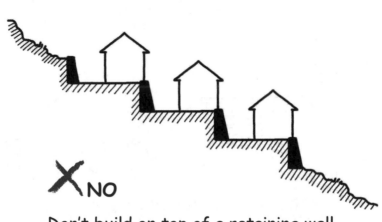

✗ NO

Don't build on top of a retaining wall.

http://dx.doi.org/10.3362/9781780449883.004

4. LAYOUT

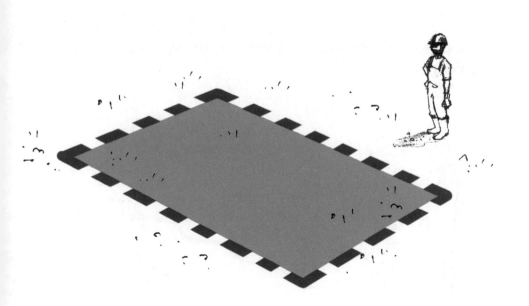

Site preparation

Remove the topsoil and the excavated material, and place it in two (or more) different heaps, away from the excavated area.

Check whether the ground is level by using a transparent hose filled with water.

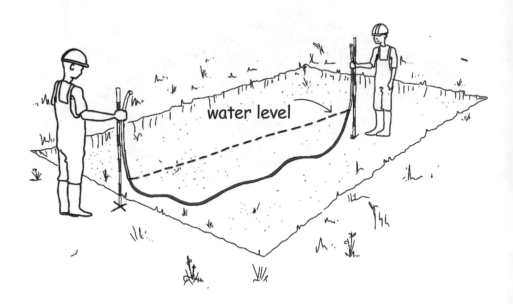

water level

Tracing a right angle (3:4:5)

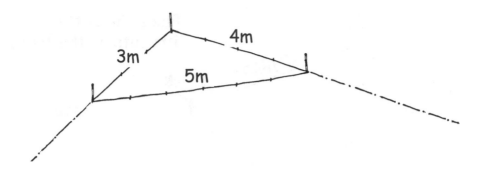

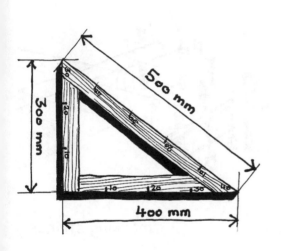

3	4	5
30 cm	40 cm	50 cm
60 cm	80 cm	100 cm
90 cm	120 cm	150 cm
1,5 m	2 m	2,5 m
2,1 m	2,8 m	3,5 m
3 m	4 m	5 m
3 ft	4 ft	5 ft
6 ft	8 ft	10 ft
9 ft	12 ft	15 ft

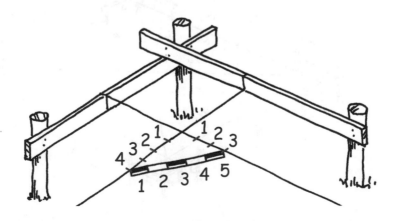

25

Layout

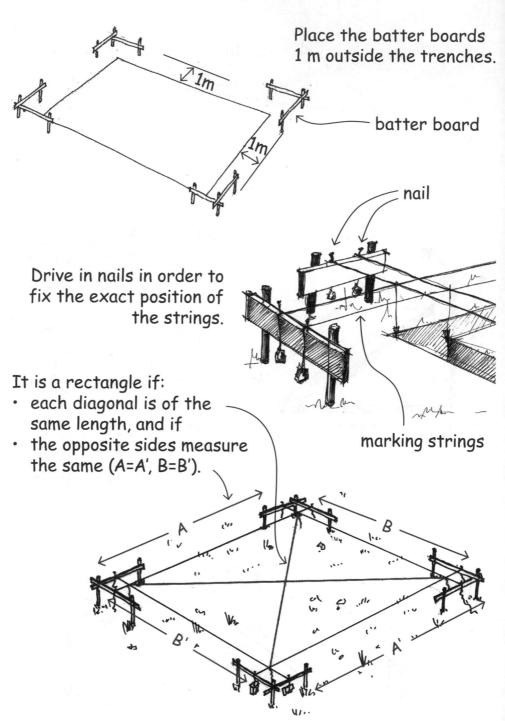

Place the batter boards 1 m outside the trenches.

batter board

nail

Drive in nails in order to fix the exact position of the strings.

marking strings

It is a rectangle if:
* each diagonal is of the same length, and if
* the opposite sides measure the same (A=A', B=B').

A

B

B'

A'

5. STONE FOUNDATION

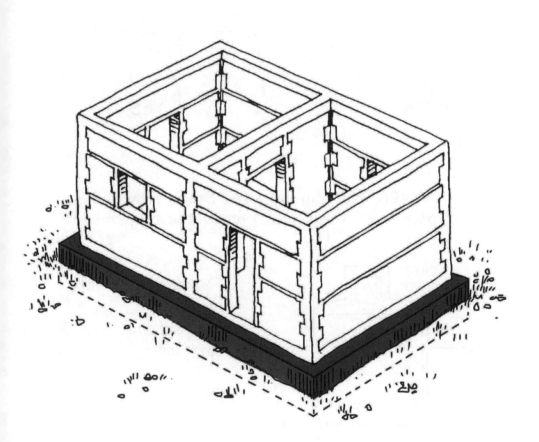

Excavation

Place the soil you have dug up at a minimum of
60 cm away from the trenches,
to avoid its falling back into the excavation.

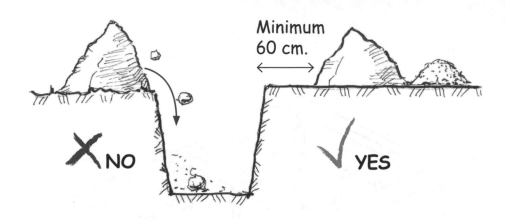

Minimum
60 cm.

✗ NO   ✓ YES

**WARNING: dig until you reach firm soil
and then build the foundation to the proper width.**

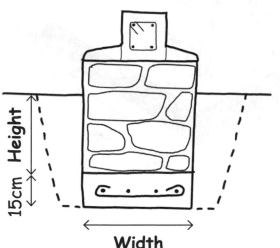

15cm Height

Width

Foundation height:
hard soil: min 30 cm
rammed soil: min 50 cm
soft soil: min 80 cm

Foundation width:
hard soil: 40 cm
rammed soil: 60 cm
soft soil: 70 cm

Foundation dimensions

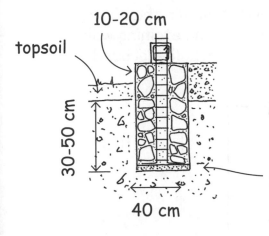

10-20 cm

topsoil

30-50 cm

40 cm

5 cm lean concrete

Hard soil
height: 30-50 cm
width: 40 cm

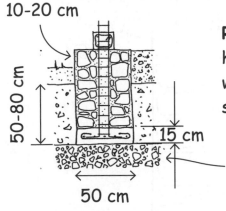

10-20 cm

50-80 cm

15 cm

50 cm

compacted soil

Rammed soil
height: 50-80 cm
width: 50 cm
strip footing: 50 cm

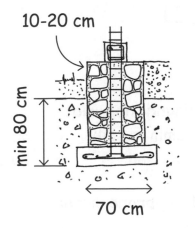

10-20 cm

min 80 cm

70 cm

Soft soil
height: min 80 cm
width: 50 cm
strip footing: 70 cm

Warning!
Height above the ground: maximum 20 cm

Special foundations

If the part above ground is higher than 20 cm,
then the foundation acts as a retaining wall.
Do not exceed 40 cm above the ground.

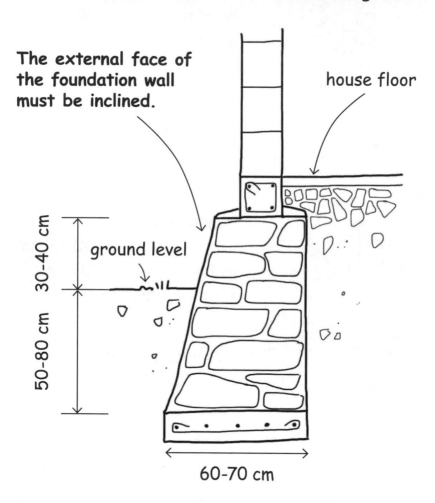

The external face of the foundation wall must be inclined.

house floor

30–40 cm

ground level

50–80 cm

60–70 cm

Foundation height:
rammed soil: min 50 cm
soft soil: min 80 cm

Foundation width:
rammed soil: min 60 cm
soft soil: min 70 cm

Avoid building in a flood-prone area!

Stepped foundations

If you build on a slope, the foundation must be stepped, **keeping the bottom of the trench always horizontal.**

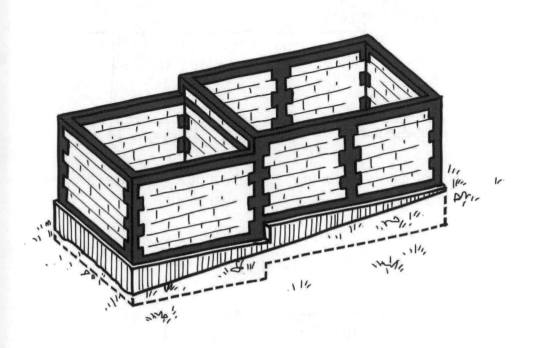

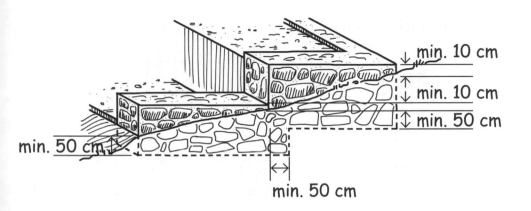

min. 10 cm

min. 10 cm

min. 50 cm

min. 50 cm

min. 50 cm

Avoid building parallel to the slope!

Stone masonry construction

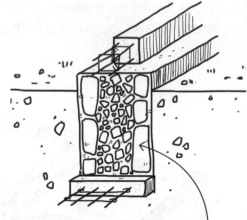

Place all the stones in a horizontal position.

Do not place the stones vertically.

 YES

 NO

Place through-stones:
Horizontally: at least every **1 m**
Vertically: at least every **50 cm**

Place through-stones

Place through-stones

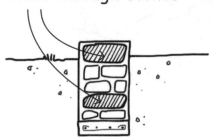

(view in section)

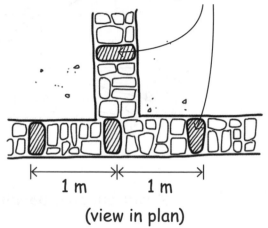

1 m   1 m

(view in plan)

Reinforced concrete strip footing

A strip footing is a must for soft soil conditions.
It is also recommended for other soil conditions.

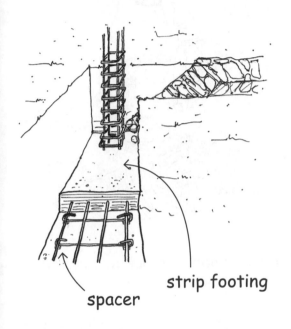

strip footing

spacer

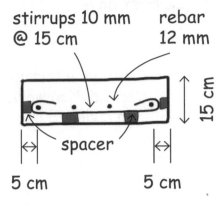

stirrups 10 mm @ 15 cm

rebar 12 mm

15 cm

spacer

5 cm   5 cm

Strip footing:
Width 40 cm = 4 rebars
Width 50 cm = 4 rebars
Width 70 cm = 5 rebars

Before pouring the concrete, make sure the reinforcement is perfectly vertical.

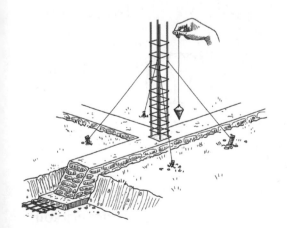

Leave a space around the reinforcement for the concrete.

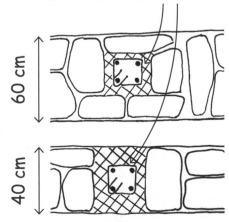

60 cm

40 cm

Curing and ground floor

Cure the foundation walls.
Wet every day for the
three first days.

Always interrupt
foundation work
on a sloped line.

Build a 'drainage pad' under the floor to block ascending humidity.

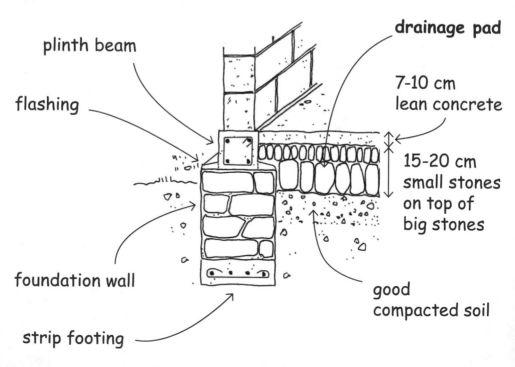

plinth beam

flashing

foundation wall

strip footing

drainage pad

7-10 cm
lean concrete

15-20 cm
small stones
on top of
big stones

good
compacted soil

Placing sewage pipes

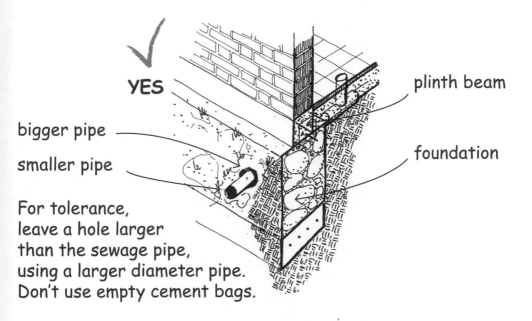

✓

YES

bigger pipe

smaller pipe

plinth beam

foundation

For tolerance,
leave a hole larger
than the sewage pipe,
using a larger diameter pipe.
Don't use empty cement bags.

**The pipe must go through the
foundation, under the plinth beam.**

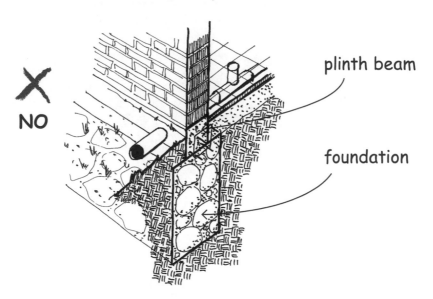

✗

NO

plinth beam

foundation

The pipe must not go through the plinth beam.

http://dx.doi.org/10.3362/9781780449883.006

6. REINFORCED CONCRETE TIES

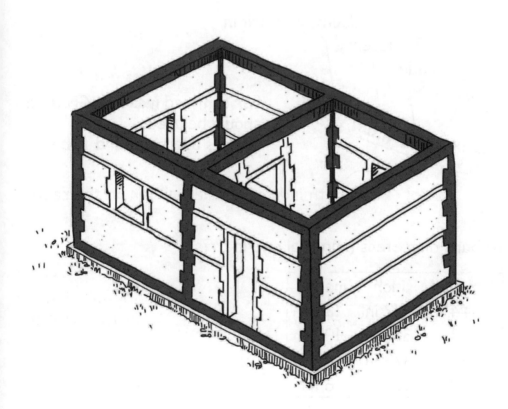

Types of steel rebars

Use ribbed steel for all rebars.

Smooth bars may only
be used for stirrups.

NO

Do not use second-hand
rebars.

Country of origin

Producer

Grade

Diameter

For confined masonry **Grade 60**
should be used.
Always use **standard rebars.**

Strength indication are written on the rebar.

Rebars diameters (imperial and metric):

imperial	inch	metric
#4	1/2 in.	12 mm
#3	3/8 in.	10 mm
-	1/3 in.	8 mm
#2	1/4 in.	6 mm

stirrups:
min Ø 6 mm
better Ø 8 mm

rebars:
min Ø 10 mm
better Ø 12 mm

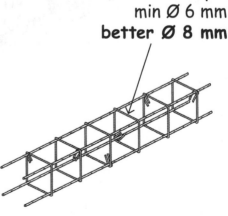

Rebar dimensions for vertical and horizontal ties

Stirrups

Bend stirrup ends at 45°.

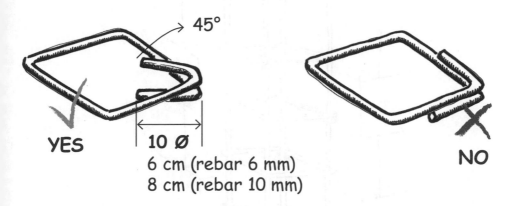

45°

YES

10 Ø
6 cm (rebar 6 mm)
8 cm (rebar 10 mm)

NO

If stirrups are not bent at 45°, they will open during an earthquake.

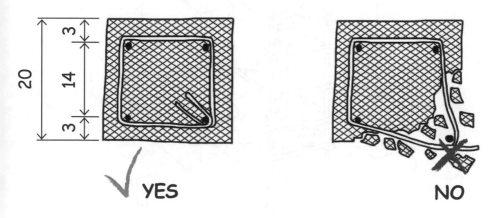

20   3   14   3

YES   NO

Possible stirrup types:

Alternate stirrup positions

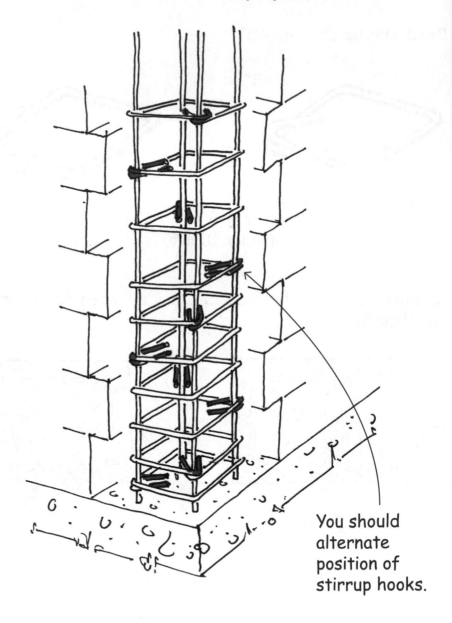

You should alternate position of stirrup hooks.

Stirrup spacing

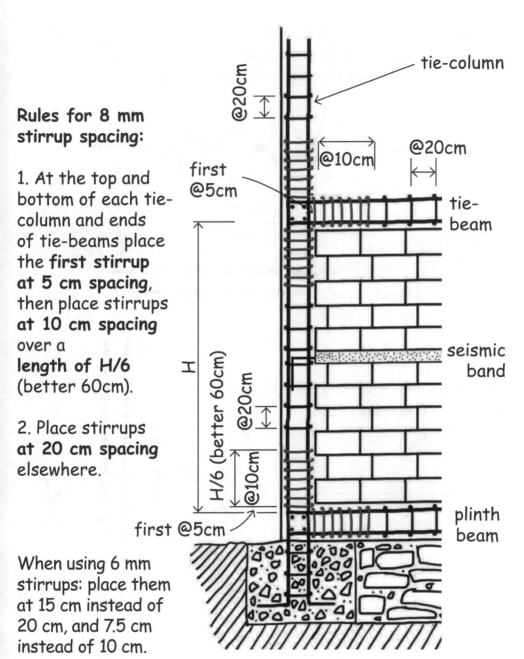

Rules for 8 mm stirrup spacing:

1. At the top and bottom of each tie-column and ends of tie-beams place the **first stirrup at 5 cm spacing**, then place stirrups **at 10 cm spacing** over a **length of H/6** (better 60cm).

2. Place stirrups **at 20 cm spacing** elsewhere.

When using 6 mm stirrups: place them at 15 cm instead of 20 cm, and 7.5 cm instead of 10 cm.

@20cm

tie-column

@20cm

@10cm

first @5cm

tie-beam

H

H/6 (better 60cm)

@20cm

@10cm

seismic band

first @5cm

plinth beam

Lap length

The concrete keeps the rebars together like tight fists:
the more fists we have (longer overlap)
the stronger the connection.

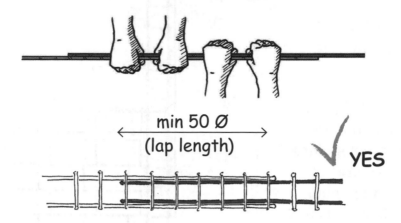

min 50 Ø
(lap length)

✓ **YES**

Tie wires only hold the rebars in place. They don't add strength to the connections! ❗

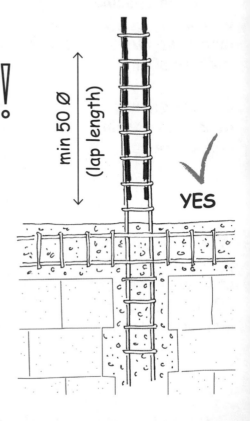

min 50 Ø
(lap length)

✓ **YES**

> **Lap length:**
> (overlapping)
> **50 x Ø**
> (50 times the diameter)
>
> for 10 mm rebar = 50 cm
> for 12 mm rebar = 60 cm

Tie-beam: T-connection

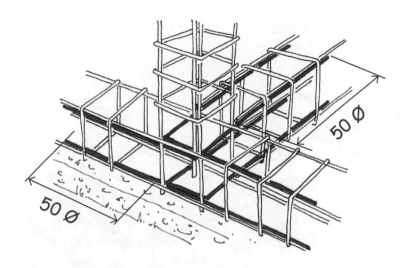

Always: extend hooked bars from the inside to the outside.

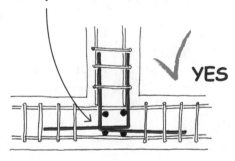

✓ YES

Lap length:
(overlapping)
50 x Ø
(50 times the diameter)

for 10 mm rebar = 50 cm
for 12 mm rebar = 60 cm

Connection with
straight bars.

Connection around
the inner corner.

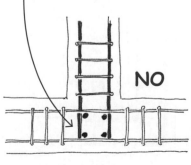

NO ✗ NO

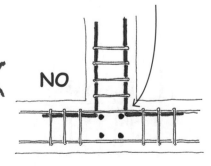

Tie-beam: L-connection

Rebars must cross like the fingers of a hand.

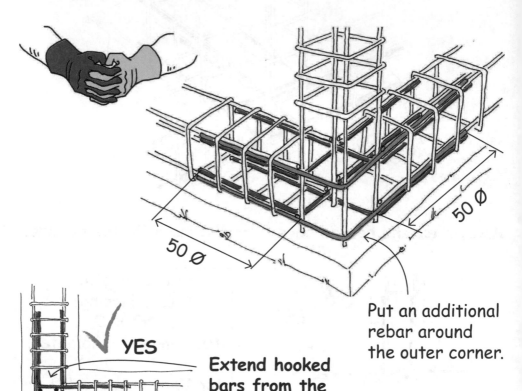

50 Ø

50 Ø

Put an additional rebar around the outer corner.

YES

Extend hooked bars from the inside to the outside.

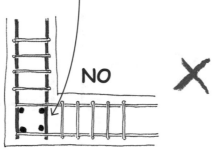

Connection with straight bars.

NO

Hooked bars from inside to inside.

NO

Tie-beam to tie-column connection

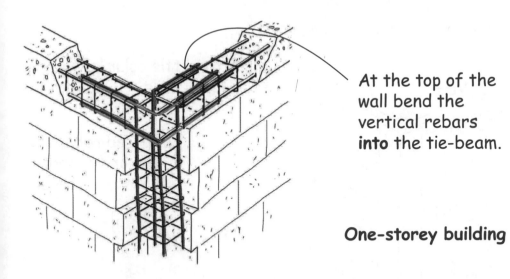

At the top of the wall bend the vertical rebars **into** the tie-beam.

One-storey building

If you plan to build an upper floor (1st floor) in the future, leave 90 cm in order to create little columns which can be used to fix guard rails.

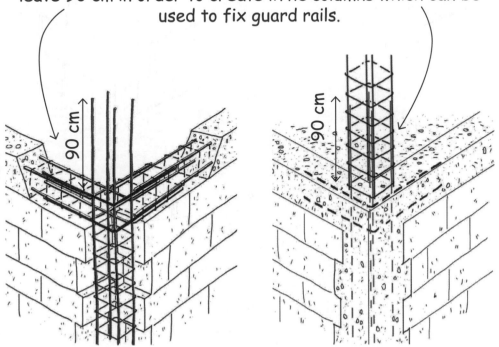

90 cm

90 cm

Two-storey building

Protection of rebar ends

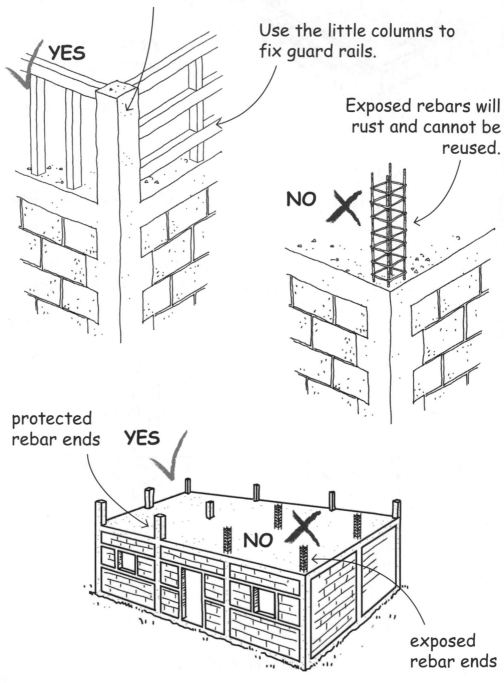

Protect rebars with lean concrete.

YES

Use the little columns to fix guard rails.

Exposed rebars will rust and cannot be reused.

NO

protected rebar ends

YES

NO

exposed rebar ends

7. FORMWORK

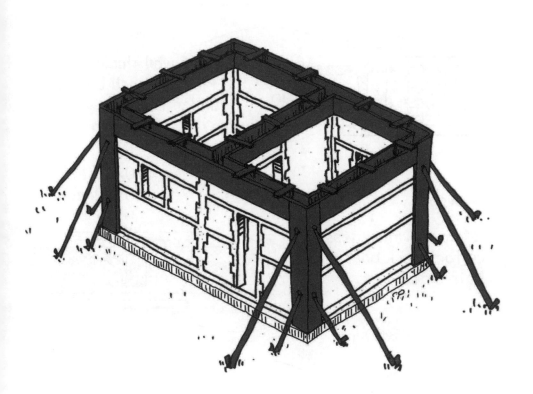

Formwork for ties

Block walls:

20cm wall thickness: place formwork boards on both sides.

Tie = 20x20cm

20 cm

Tie = 20x20cm

15 cm

15cm wall thickness: Place a 1 inch board under the formwork board.

Brick walls:

15-24cm wall thickness: place formwork boards on both sides.

Tie = 20x20cm
min 15x20cm

min 15 cm

Tie = 15x20cm

min 11 cm

15cm wall thickness: Place a 1 inch board under the formwork board.

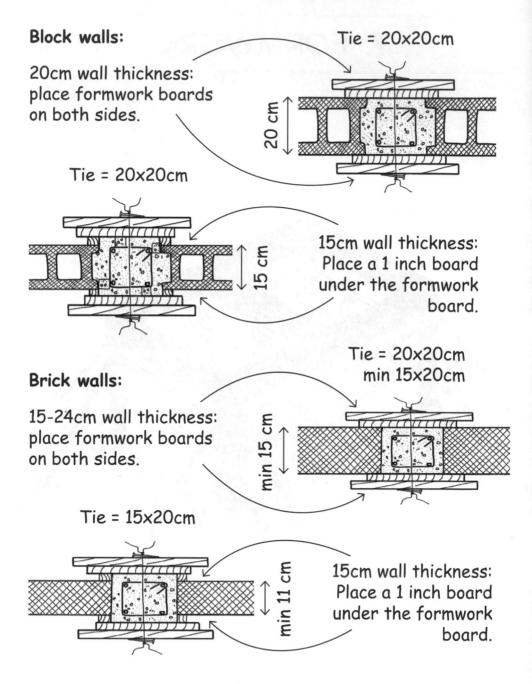

**Sizes of tie-columns and tie-beams:
20 x 20 cm recommended / 15 x 20 cm minimum**

Vertical formwork

Formwork fixed with wires

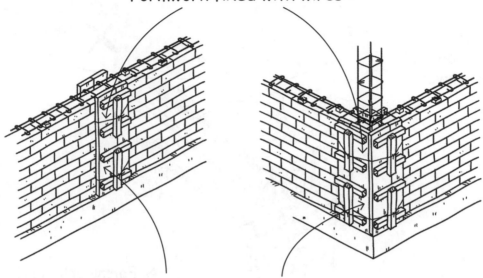

Attention: with this type of formwork wait until the masonry is solid or the wires will move the bricks.

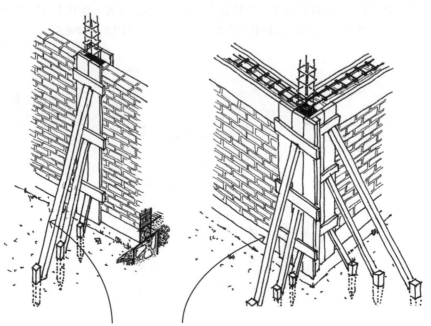

Formwork must be well braced.

Horizontal formwork

YES ✓

Use wood planks to connect formwork.

NO ✗ Don't use tie wire.

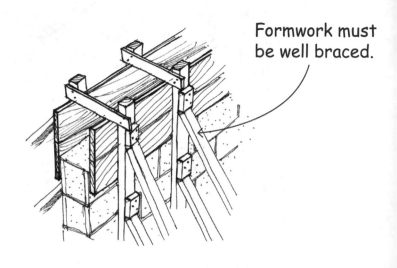

Formwork must be well fastened.

Using small planks to keep the formwork apart ensures more precision and stability than wires.

Formwork must be well braced.

Spacers – 1

Spacers are very important: they ensure that the rebars remain in the right place and are well covered by concrete.

Don't use stones to fix the rebars, use spacers instead.

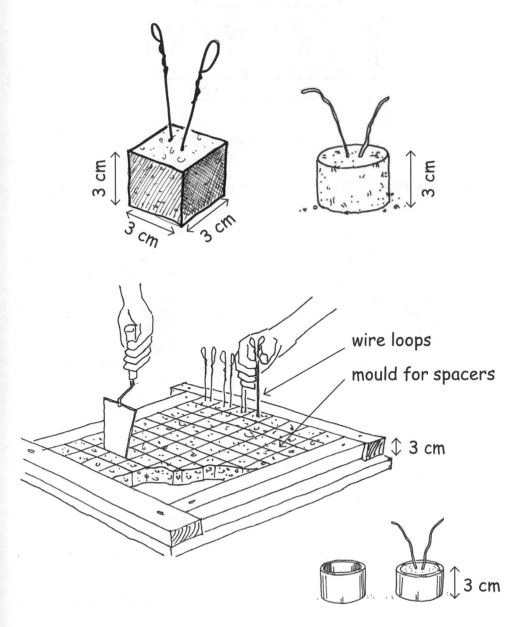

3 cm

3 cm

3 cm

3 cm

3 cm

wire loops

mould for spacers

↕ 3 cm

↕ 3 cm

51

Spacers – 2

Add spacers on all sides
to avoid rebars touching the formwork.

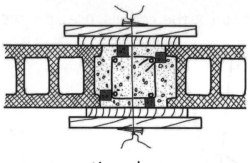

tie-column

Alternate the position of the spacers around the stirrups.

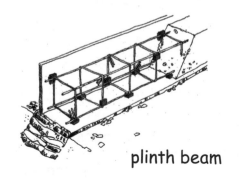

plinth beam

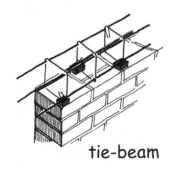

tie-beam

reinforced concrete slab

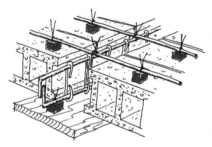

joist and pan slab

http://dx.doi.org/10.3362/9781780449883.008

8. CONCRETE

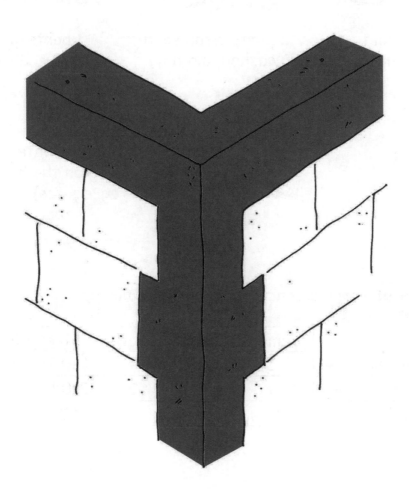

Concrete mix (1:2:3)

| 1 part cement | 2 parts clean sand (washed and dry) | 3 parts gravel (max. 18mm) |

3/4 part clean water (depending on the humidity of the agregates)

Table of various concrete mixes (by volume):

	Cement	Sand	Gravel	Strength
minimum →	1	2	4	180 kg/cm²
standard →	1	2	3	210 kg/cm² ✓
ideal →	1.5	2	3	240 kg/cm²

Note:
Concrete with a strength of 210 kg/cm2 corresponds more or less to 350 kg of cement per cubic meter of aggregates.

Mixing concrete

Mixing the concrete by hand:

1. Make a pile with the gravel, the sand and the cement but without water.

2. Mix the pile without water and move it twice with a shovel.

3. Add the water and mix again. **Only add the water at the end.**

Mixing with a concrete mixer:

1. Add half of the water and 1 part of gravel, mix 1 minute.

2. Add the cement and the rest of the agregates.

3. Add rest of the water slowly, mix 3-4 minutes (not more).

Always use the concrete within 90 minutes of mixing!

Concrete test

QUICK TEST:
Take a handful of concrete. If you
can form a nice ball, the concrete
is perfect. If the concrete leaks
through your fingers, it is too wet.

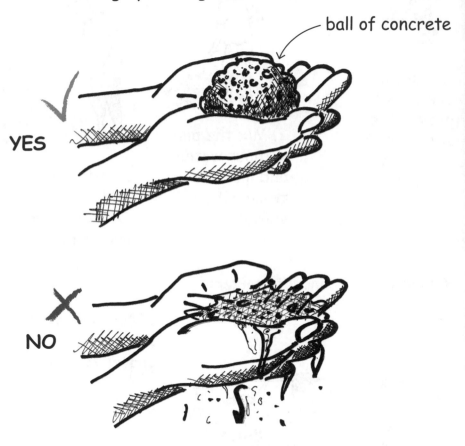

ball of concrete

YES

NO

Concrete must be **used in less than 90 minutes**.
Never 'refresh' dried concrete by adding water.
Don't mix too much concrete at a time.

Slump test

SLUMP TEST PROCEDURE:

Use a standard steel cone:

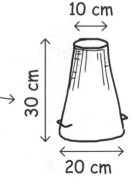

10 cm

30 cm

20 cm

1. Fill cone in 3 equal layers.

2. Tamp down each layer 25 times with a rod (rebar).

3. Lift the cone vertically and place next to the slump.

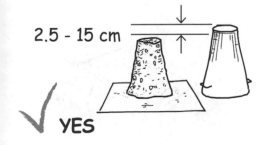

2.5 - 15 cm

✓ YES

more than 15 cm

NO ✗

Interpretation of results:

Workability	mm	Use
low	25-50	foundations with little reinforcement
medium	50-100	for compacted and vibrated concrete
high	100-150	parts with very congested reinforcements and narrow structural elements

Pouring and compacting concrete

Pour the concrete in layers of 30 to 60 cm and compact well with a rod and hammer, or better, with a vibrating needle.

Never add water to make the concrete more liquid and 'flow down better'.

Roughen up the top surface of the plinth beam to increase bonding of the mortar for the wall.

Compacting with a vibrating needle

By compacting the concrete with a vibrator needle, trapped air will rise to the surface in the form of air bubbles.

1. Insert the vibrating needle 10 cm into the previous layer.

2. Leave the needle for not more than 5 to 15 seconds to avoid the concrete disintegrating.

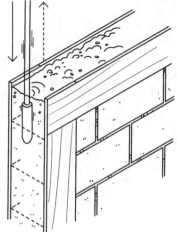

3. Lift the needle slowly (air bubbles rise at a speed of 2.5 to 7.5 cm per second).

4. Don't touch the reinforcement steel while vibrating.

5. Don't use the needle to move the concrete sideways.

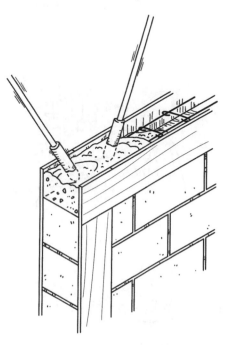

Advance at regular intervals following the action radius of the needle.

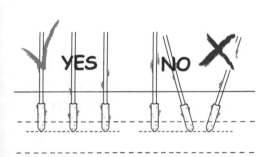

YES

NO

Curing the concrete elements

Concrete needs water to harden.

After placing concrete,
cure the concrete by
wetting the formwork
three times a day
for three days.
Remove formwork only
after three days.

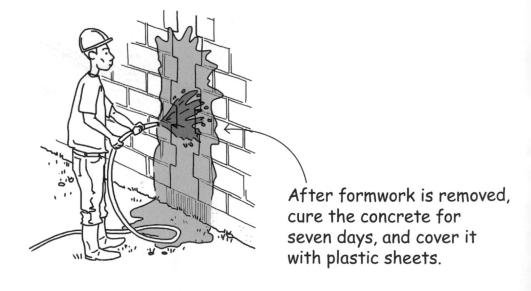

After formwork is removed,
cure the concrete for
seven days, and cover it
with plastic sheets.

Ensure good-quality concrete

Exposed rebars will rust.

Poor compaction: the concrete is weakened.

http://dx.doi.org/10.3362/9781780449883.009

9. BRICKS AND BLOCKS

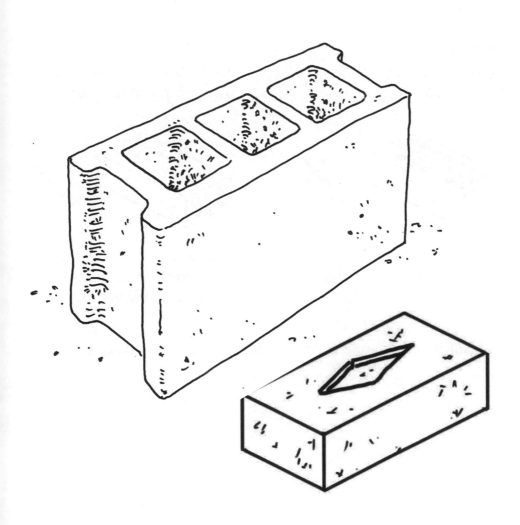

Which clay bricks to use

Frog

YES ✓

Best brick:
solid burnt-clay brick
with frogs.

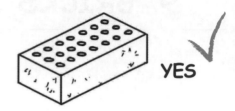

YES ✓

Good brick:
vertical holes less than
50% of surface area.

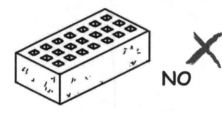

NO ✗

Bad brick:
vertical holes more than
50% of surface area.

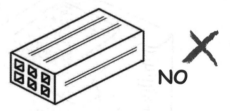

NO ✗

Bad brick:
with horizontal holes
(cannot carry weight).

Solid bricks are better than multiperforated ones.

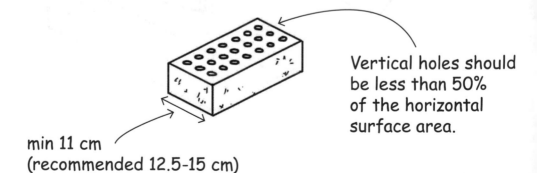

Vertical holes should
be less than 50%
of the horizontal
surface area.

min 11 cm
(recommended 12.5-15 cm)

Note: we recommend using 10MPa bricks.

Brick test

Visual test:

1. regular in form

2. uniform colour

3. not warped

4. no visible flaws or lumps

Physical test:

1. Bricks cannot be easily scratched by a knife.

2. Resists the **'3 point test'**: Person stands on a brick spanning two other bricks.

3. Bricks must give a ringing sound when struck against each other.

Which concrete blocks to use

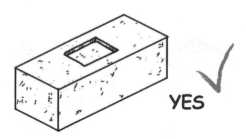

YES

Best block:
15-20 cm thick,
solid block.

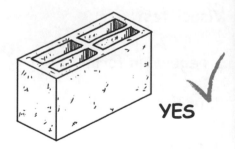

YES

Best block:
15-20 cm thick,
with 4 holes.

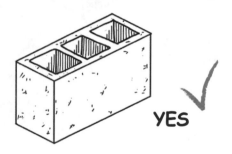

YES

Satisfactory block:
18-20 cm thick,
with 3 holes.

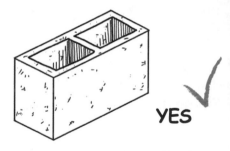

YES

Only if excellent quality:
20 cm thick,
with 2 holes.

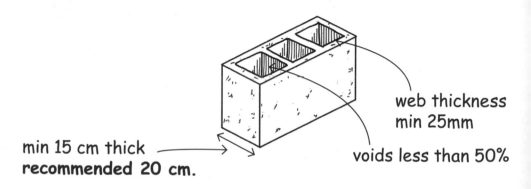

web thickness
min 25mm

min 15 cm thick
recommended 20 cm.

voids less than 50%

Note: we recommend using 10MPa blocks.

Block test

Test blocks before buying them

Drop five blocks from 1.5 m height on a hard surface (concrete surface).

YES ✓

NO ✗

Acceptable quality:
(less than one broken)

Bad quality: don't buy
(if more than one broken)

Check if blocks were
cured in the shade.

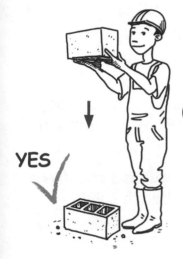

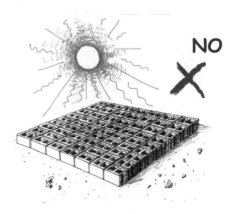

YES ✓

NO ✗

Stored in the shade: good

YES ✓

Stored under plastic sheets: good

Blocks that dry in
the sun: very bad

Concrete mix for blocks (1:4:3)

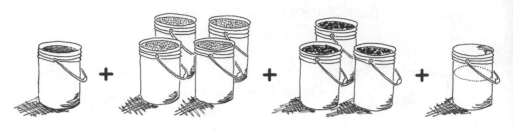

| 1 part cement | + | 4 parts clean sand | + | 3 parts gravel (8-10mm) | + | 3/4 part clean water |

Sand should be crushed, washed and dried.
Do not use marine beach sand.

1. Make a pile with the gravel, the sand and the cement but without water.

2. Mix the pile without water and move it twice with a shovel.

3. Add water and mix again.

Add water only at the end.

68

Making the blocks

Wait eight days before using the blocks.

Fill the molds with the mixture.

If possible use a vibrating machine

To compact the concrete, hit the mold with a shovel and a hammer.

Cover the blocks with plastic sheets immediately.

Store the blocks in the shade.

Cure the blocks three times a day for **minimum seven days** and cover with plastic sheets.

http://dx.doi.org/10.3362/9781780449883.010

10. MASONRY WALLS

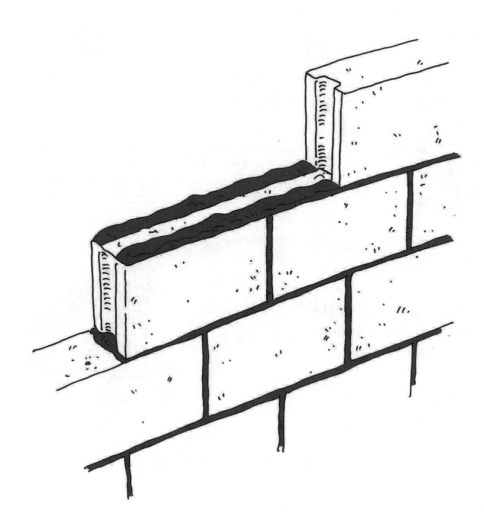

Cement mortar mix (1:4)

Mix the mortar:

 +  +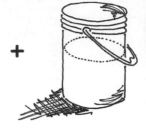

| 1 part cement | 4 parts clean sand (washed and dry) | 3/4 part clean water |

Use 1:3 mix ratio for 15cm or less wall thickness

1. Make a pile with the sand and the cement but without water.

2. Mix the pile without water and move it twice with a shovel.

3. Add the water and mix again.

Add the water only at the end.

Cement-lime mortars

Cement-lime mortar
has lower compressive strength than simple cement mortar
but offers a better workability, higher elasticity,
and is more economical.

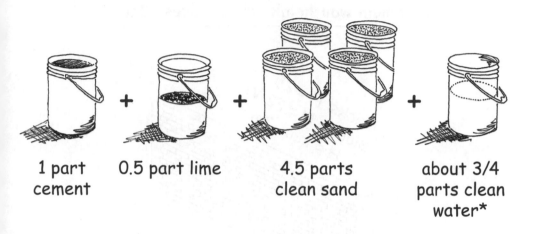

| 1 part cement | 0.5 part lime | 4.5 parts clean sand | about 3/4 parts clean water* |

* enough water to get a good workability

Recommended mortar mix proportions:

	Cement	Lime	Sand
ideal ⟶	1	0.5	4.5
	1	1	6
minimum ⟶	1	2	9

Masonry walls height

The width of masonry unit defines the wall height.

> - Bricks: maximum height = 25 x wall width
> - Blocks: maximum height = 22 x wall width
> - **Maximum wall height in all cases: 3m**

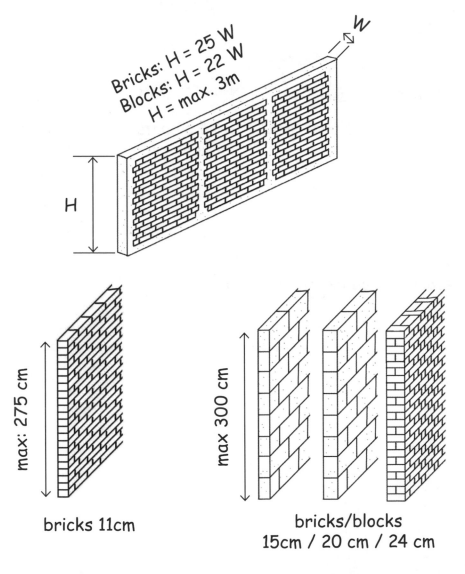

bricks 11cm

bricks/blocks
15cm / 20 cm / 24 cm

Masonry bonds

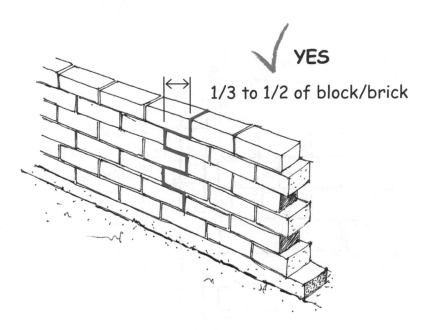

✓ **YES**

1/3 to 1/2 of block/brick

Solid wall = running bond –
vertical joints are not continous.

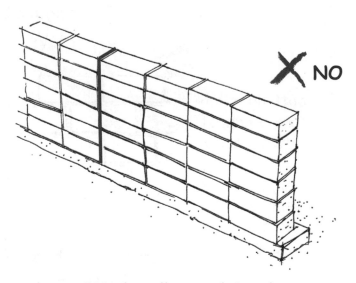

✗ **NO**

Weak wall = stack bond –
vertical joints are continuous.

Toothing

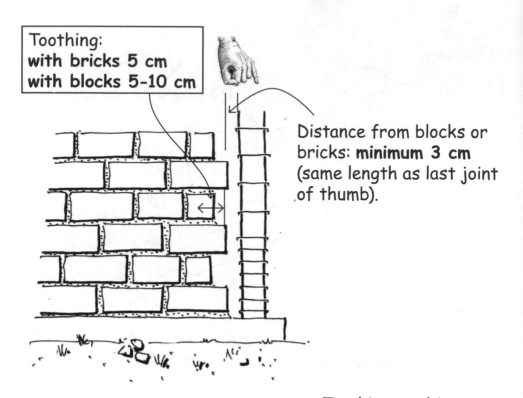

Toothing:
with bricks 5 cm
with blocks 5-10 cm

Distance from blocks or bricks: **minimum 3 cm** (same length as last joint of thumb).

Toothing too big.

If toothing is too big, concrete cannot penetrate properly.

Also, the weight of the concrete may cause bricks or blocks to break off during the pouring process.

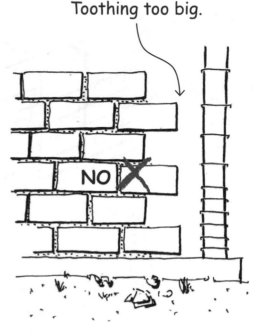

NO

Preparing the masonry units

... or ...

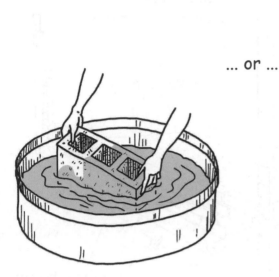

Soak the blocks in water for a while...

... water them with a brush before use.

... or ...

... water all blocks together.

Good masonry practice – 1

Use a plank as guide to ensure the wall is in plumb and straight.

Place blocks one course at a time

level string

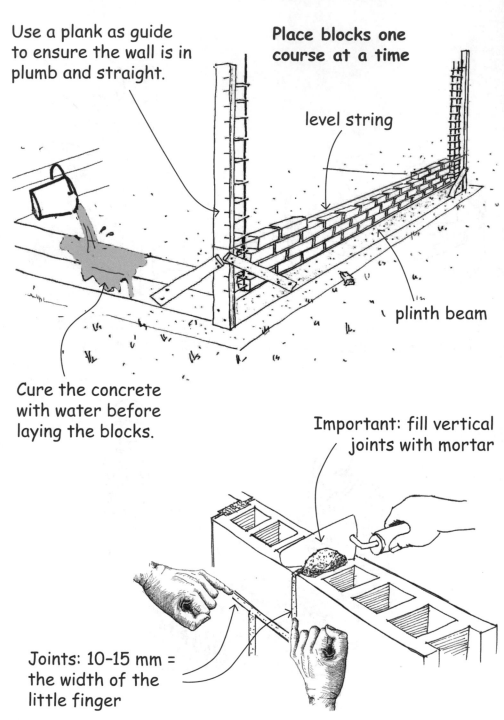

plinth beam

Cure the concrete with water before laying the blocks.

Important: fill vertical joints with mortar

Joints: 10–15 mm = the width of the little finger

Good masonry practice – 2

Don't build more than six courses of masonry per day. And then add a seismic band if needed.

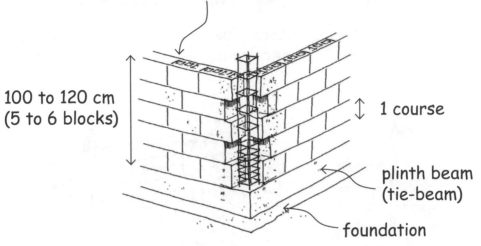

100 to 120 cm
(5 to 6 blocks)

1 course

plinth beam
(tie-beam)

foundation

Protect the wall in warm weather:
mortar must not dry out in the sun.

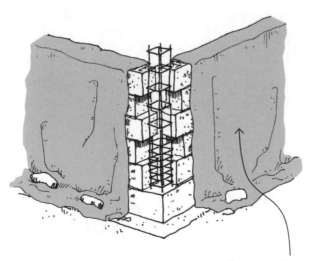

Keep wall moist by pouring water on it three times a day
for seven days and/or by covering them with a plastic
sheet for seven days.

Placing pipes in walls

✓ YES

Place pipes in
block holes.

✓ YES

Place pipes in
service duct.

✗ NO

Don't place pipes
in walls or in ties.

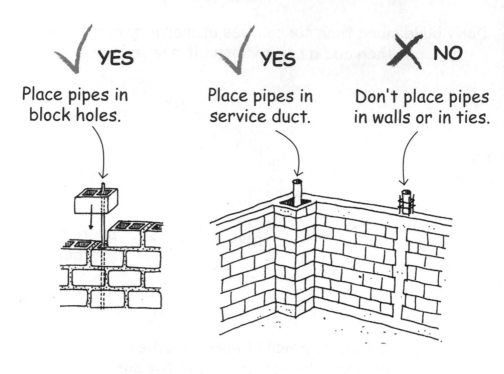

Dont break masonry
to insert pipes.

The best ways to
place pipes is on top
of the plaster.

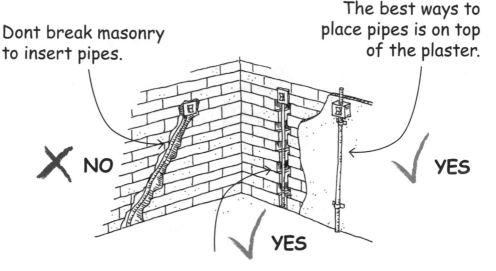

✗ NO

✓ YES

✓ YES

Leave a space in the wall for electrical pipes.
Once the pipes placed it will be filled with mortar.

http://dx.doi.org/10.3362/9781780449883.011

11. SEISMIC REINFORCEMENTS

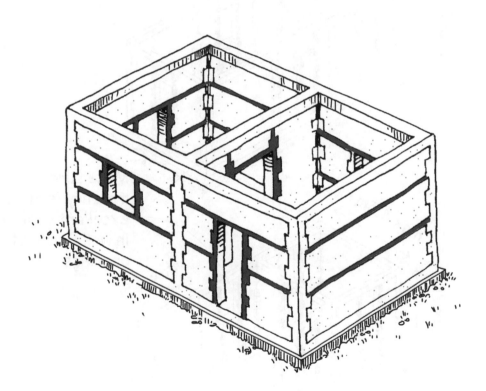

Vertical reinforcement of openings

There are two types of reinforcements of openings, vertical and horizontal. They are equally valid.
For horizontal reinforcements see p. 86.

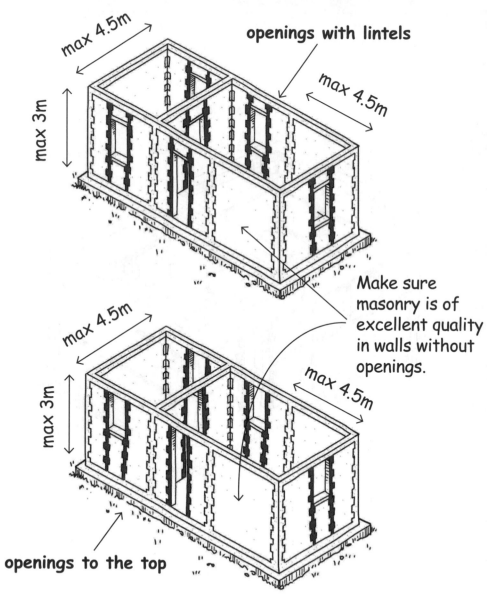

max 4.5m

openings with lintels

max 4.5m

max 3m

Make sure masonry is of excellent quality in walls without openings.

max 4.5m

max 3m

max 4.5m

openings to the top

Place a vertical reinforcement on each side of every opening.

Door reinforcement

Hook the door vertical reinforcement rebars and lap 30 cm with the tie-beam rebars, under the stirrups. Do the same with the lintel band and the vertical bands.

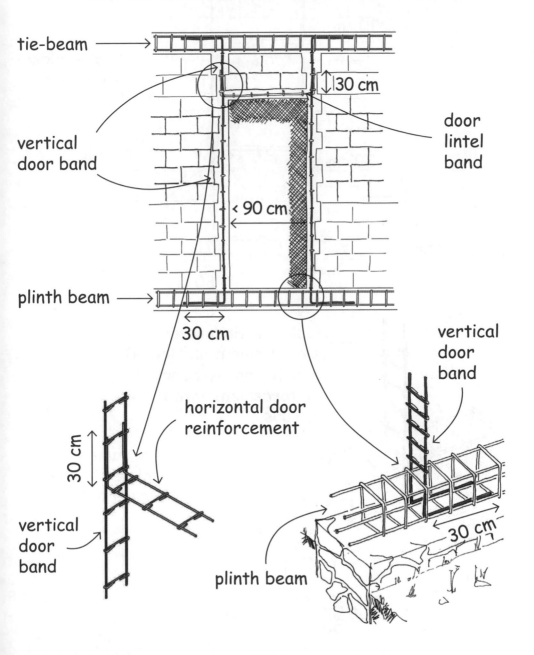

tie-beam

30 cm

vertical door band

door lintel band

< 90 cm

plinth beam

30 cm

horizontal door reinforcement

vertical door band

30 cm

vertical door band

plinth beam

30 cm

Reinforcement of small windows

For windows smaller than 90 cm.

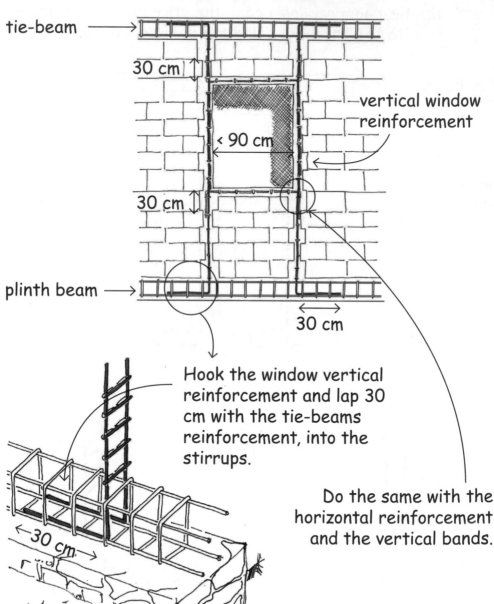

tie-beam

30 cm

< 90 cm

vertical window reinforcement

30 cm

plinth beam

30 cm

Hook the window vertical reinforcement and lap 30 cm with the tie-beams reinforcement, into the stirrups.

Do the same with the horizontal reinforcement and the vertical bands.

30 cm

Reinforcement of large windows

For windows larger than 90 cm.

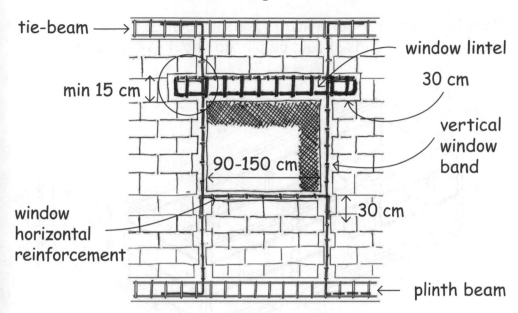

tie-beam →

window lintel

30 cm

min 15 cm

vertical
window
band

90-150 cm

window
horizontal
reinforcement

30 cm

plinth beam ←

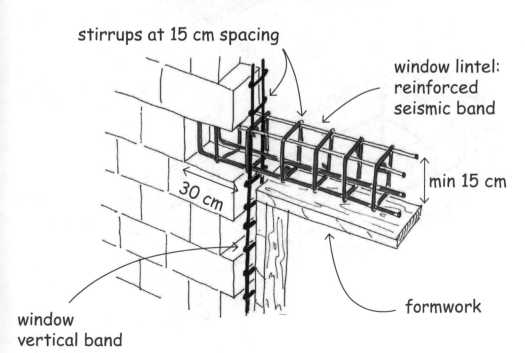

stirrups at 15 cm spacing

window lintel:
reinforced
seismic band

30 cm

min 15 cm

window
vertical band

formwork

85

Horizontal reinforcements (seismic bands)

Place horizontal reinforcements (seismic bands) below and above every opening. Bands should be placed about every 1.2 m.

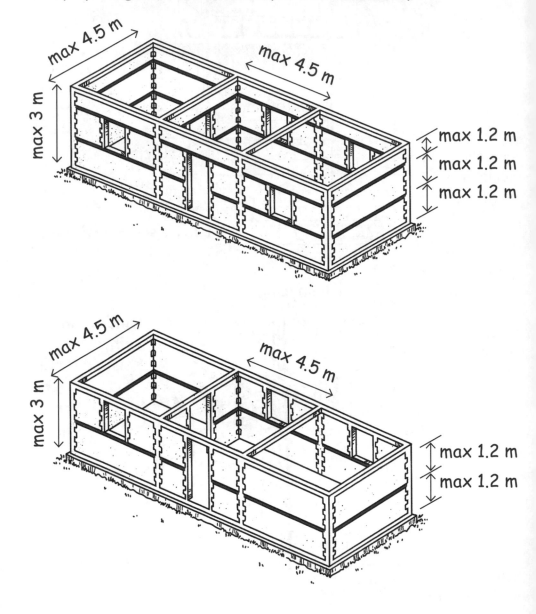

Seismic bands

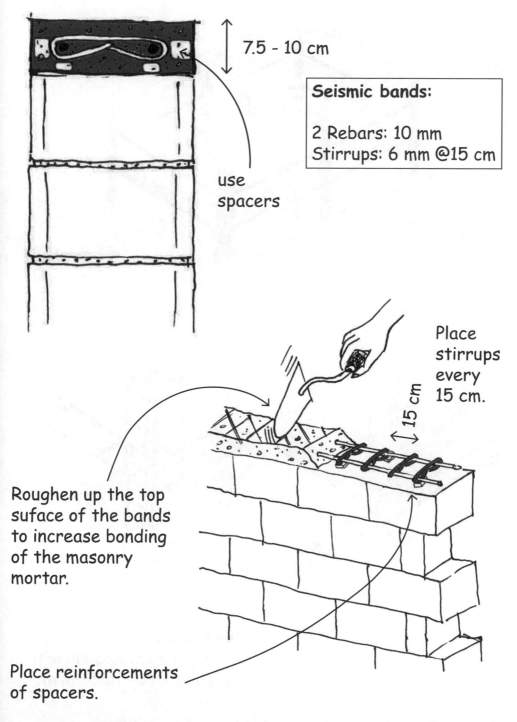

7.5 - 10 cm

Seismic bands:

2 Rebars: 10 mm
Stirrups: 6 mm @15 cm

use
spacers

Place
stirrups
every
15 cm.

15 cm

Roughen up the top
suface of the bands
to increase bonding
of the masonry
mortar.

Place reinforcements
of spacers.

Connections

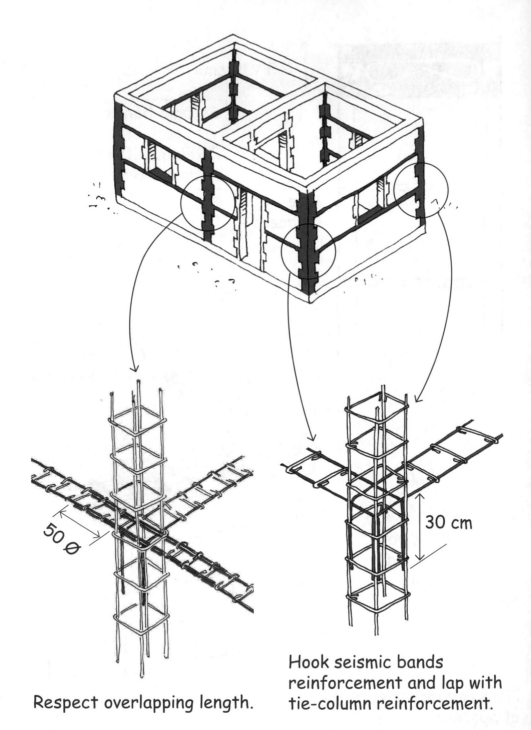

50 ⌀

30 cm

Respect overlapping length.

Hook seismic bands reinforcement and lap with tie-column reinforcement.

Reinforcement of small windows

For windows smaller than 90 cm.

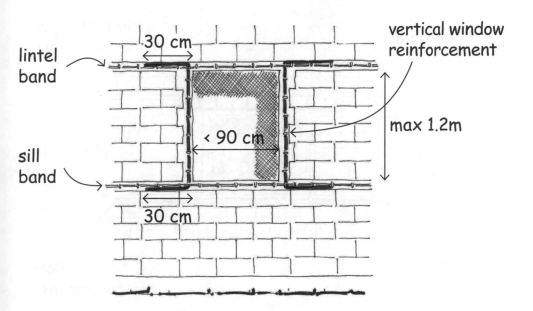

lintel band

vertical window reinforcement

30 cm

max 1.2m

sill band

‹ 90 cm

30 cm

Hook the window reinforcement and lap 30 cm with the seismic band reinforcement, into the stirrups.

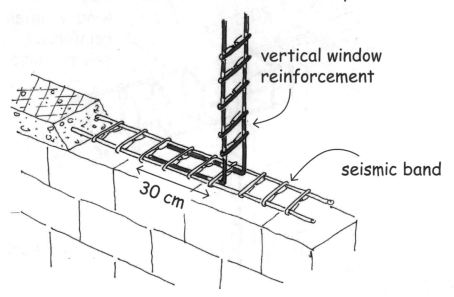

vertical window reinforcement

30 cm

seismic band

Reinforcement of large windows

For windows larger than 90 cm.

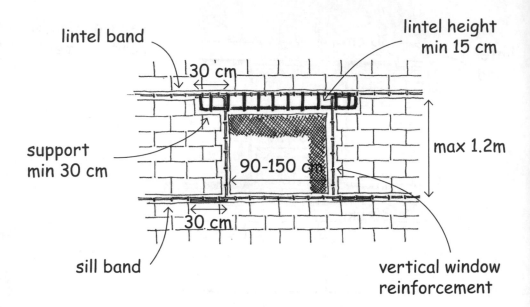

lintel band

30 cm

lintel height
min 15 cm

support
min 30 cm

90-150 cm

max 1.2m

sill band

30 cm

vertical window
reinforcement

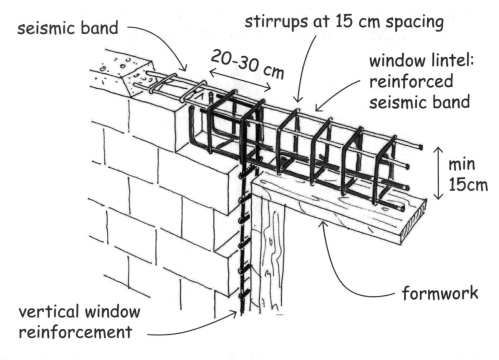

seismic band

stirrups at 15 cm spacing

20-30 cm

window lintel:
reinforced
seismic band

min
15cm

formwork

vertical window
reinforcement

Creating shear walls using vertical reinforcements

Vertical bands are 'half tie-columns' with **only two rebars.**

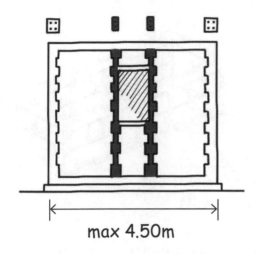

max 4.50m

Vertical bands:
(for openings)

Width: 10 cm
2 Rebars: 10 mm
Stirrups: 6 mm (@ 15cm)

If a wall between openings is required to act as a shear wall, the vertical reinforcement is identical to a tie-column with **four rebars**

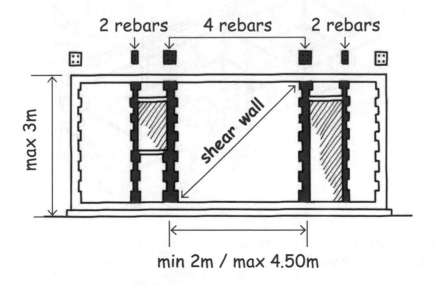

2 rebars   4 rebars   2 rebars

max 3m

shear wall

min 2m / max 4.50m

Shear walls with horizontal bands 1

In some cases it might seem impossible to provide shear walls in each facade because the owner wants too many windows.

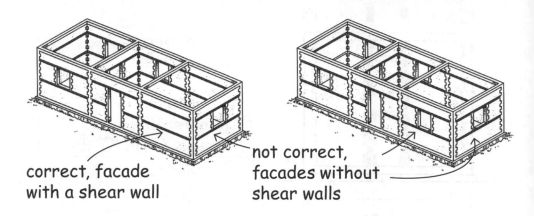

correct, facade
with a shear wall

not correct,
facades without
shear walls

In these cases shear walls are created by increasing the reinforcements on the side of some specific openings.

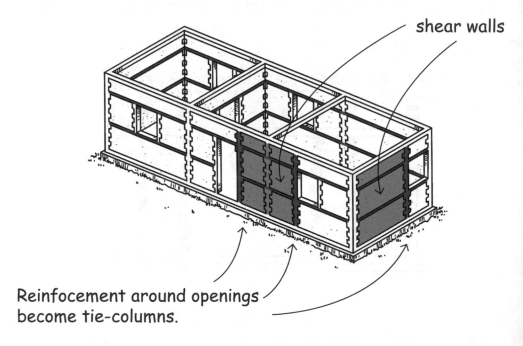

shear walls

Reinfocement around openings
become tie-columns.

Shear walls with horizontal bands 2

The vertical reinforcements of the openings (with 2 rebars) can be made like tie-columns with 4 rebars and extending them down to the plinth and up to the ring beam.

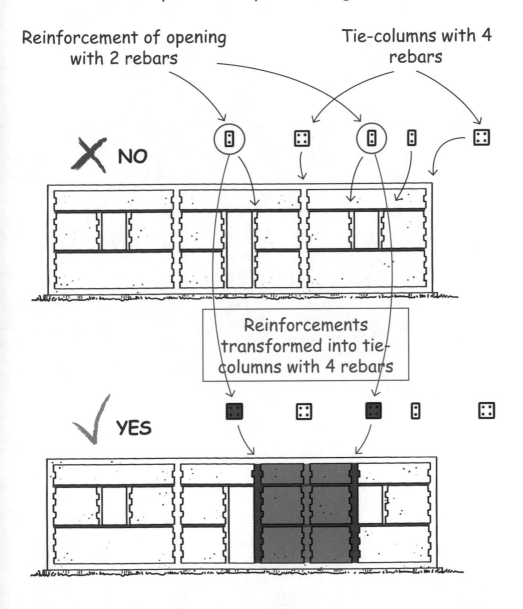

Reinforcement of opening with 2 rebars

Tie-columns with 4 rebars

X NO

Reinforcements transformed into tie-columns with 4 rebars

✓ YES

12. SLAB

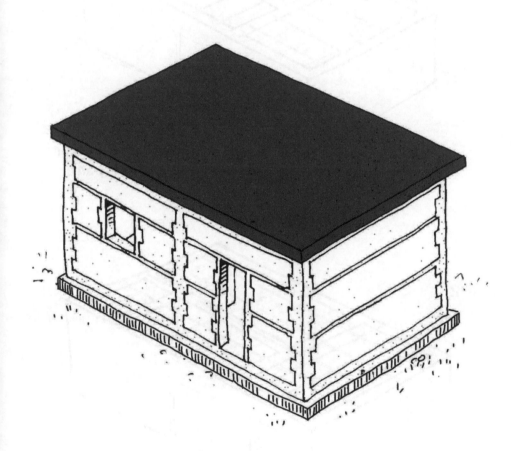

Placing of slab reinforcement

Placement of primary rebars.

Step 1

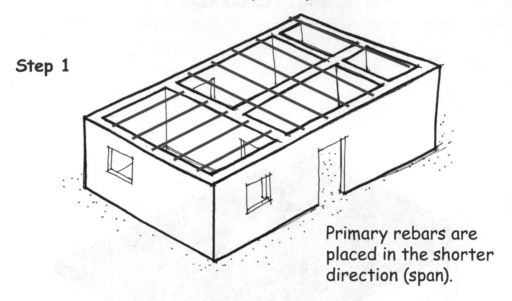

Primary rebars are placed in the shorter direction (span).

Placement of secondary rebars.

Step 2

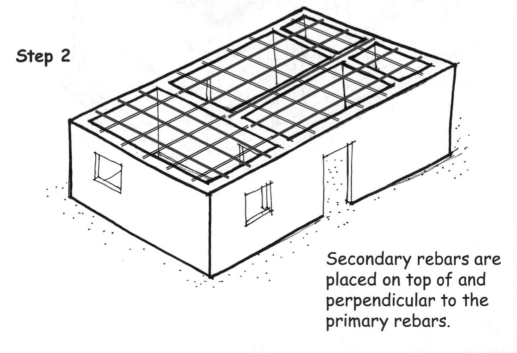

Secondary rebars are placed on top of and perpendicular to the primary rebars.

Hollow block slab: formwork

GOOD FORMWORK

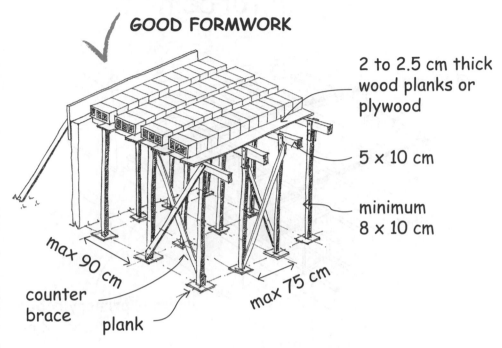

2 to 2.5 cm thick wood planks or plywood

5 x 10 cm

minimum 8 x 10 cm

max 90 cm

counter brace

plank

max 75 cm

BAD FORMWORK

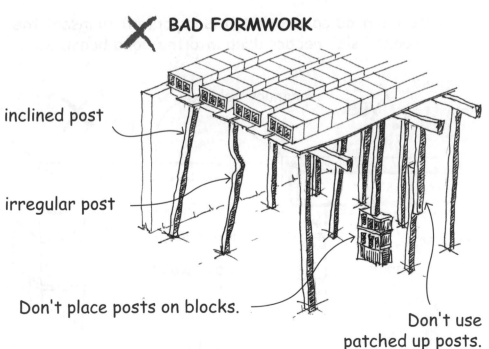

inclined post

irregular post

Don't place posts on blocks.

Don't use patched up posts.

Hollow block slab: main reinforcement

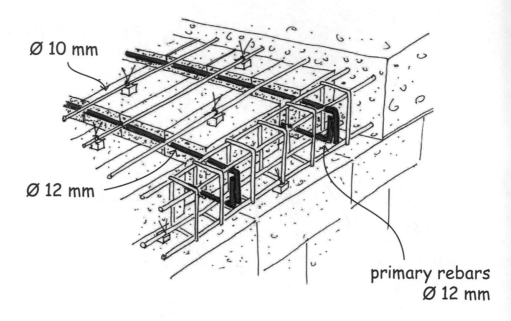

Ø 10 mm

Ø 12 mm

primary rebars
Ø 12 mm

To ensure a good connection, it is important to insert the hooked slab rebars deep into the bond beam.

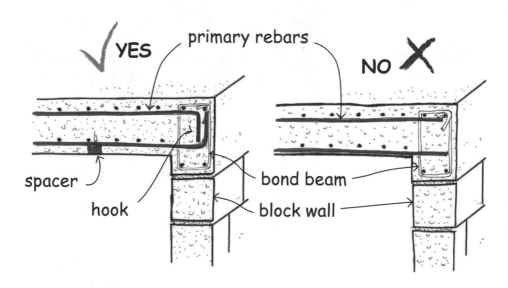

✓ YES

primary rebars

NO ✗

spacer

hook

bond beam

block wall

Hollow block slab: secondary rebars

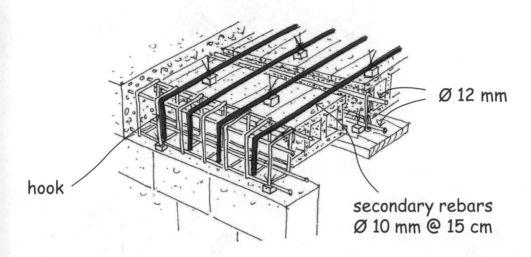

Ø 12 mm

hook

secondary rebars
Ø 10 mm @ 15 cm

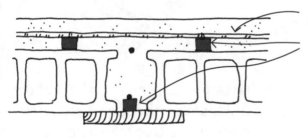

secondary rebars

spacers

Secondary rebars must be placed in the middle of the concrete covering the hollow blocks with spacers.

✓ YES

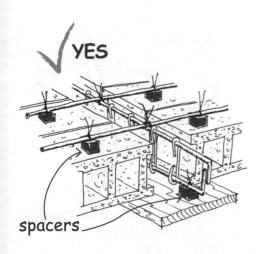

spacers

✗ NO

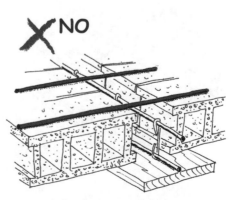

missing spacers

99

Placing pipes in hollow block slabs

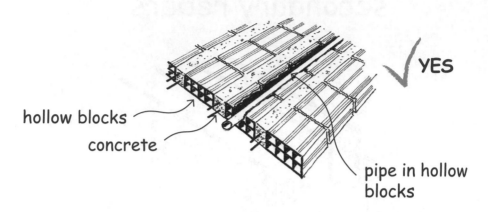

YES

hollow blocks

concrete

pipe in hollow blocks

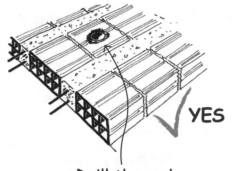

YES

Drill through hollow blocks.

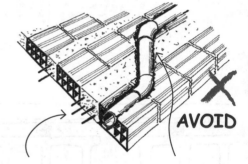

AVOID

Pass pipes through the hollow blocks and through concrete only in one spot. Reinforce joist with additional rebars.

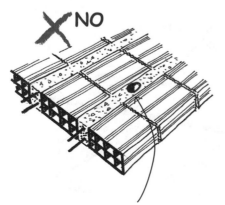

NO

Don't drill through concrete.

NO

Don't cross concrete all the way.

Preparing the slab for concrete

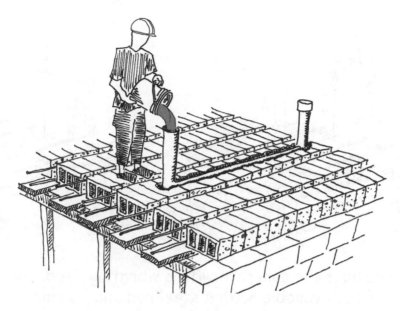

Test watertightness of the pipes by filling them with water and wait for four hours.

Water the formwork before pouring the concrete.

Pouring the concrete

Compact the concrete with a vibrating needle or, if not available, with a steel rod and hammer.

Curing the concrete: create ponds with sand or mud and fill them with water for a week.

7 days

http://dx.doi.org/10.3362/9781780449883.013

13. LIGHT ROOF

Roof shape

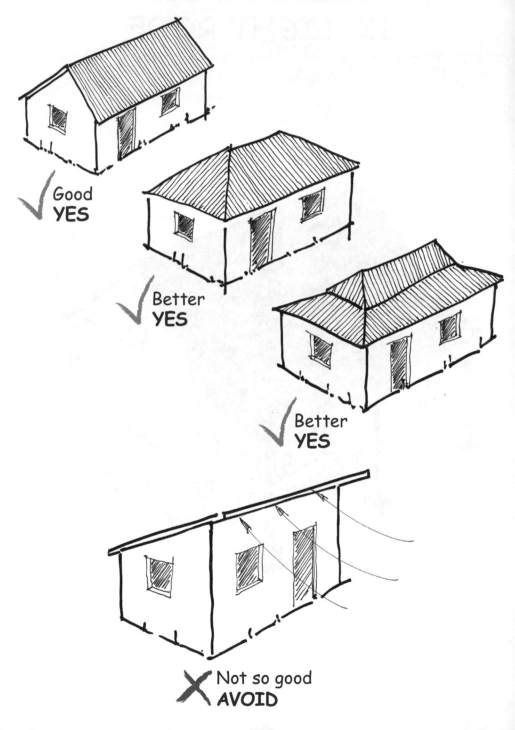

Good
YES

Better
YES

Better
YES

Not so good
AVOID

Gable wall

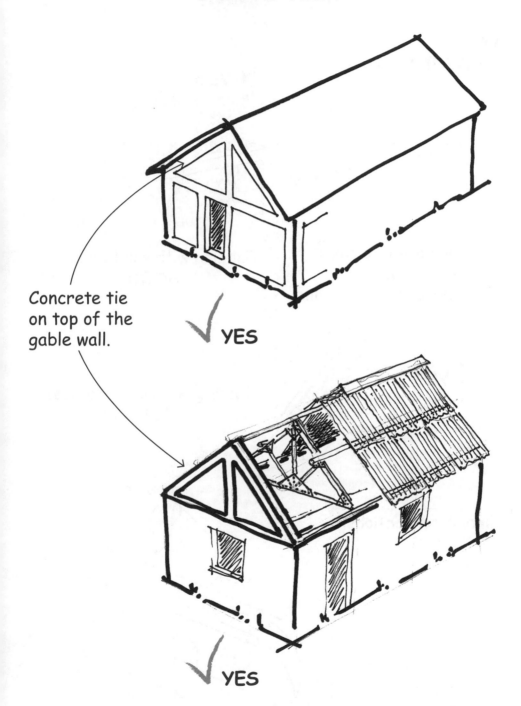

Concrete tie on top of the gable wall.

✓ YES

✓ YES

Roof structure - Trusses

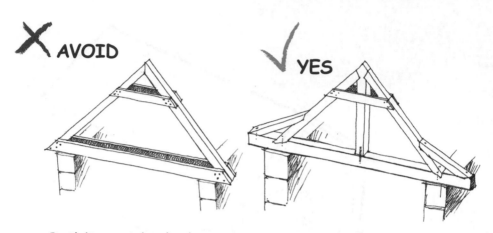

✗ AVOID

✓ YES

Building with planks:
AVOID
(not enough room for nails)

Building with solid timber:
GOOD

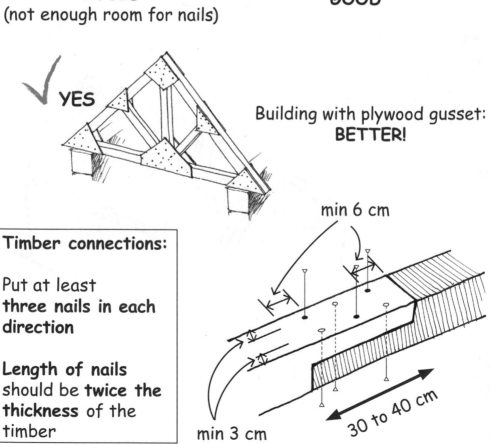

✓ YES

Building with plywood gusset:
BETTER!

Timber connections:

Put at least
three nails in each direction

Length of nails should be **twice the thickness** of the timber

min 6 cm

min 3 cm

30 to 40 cm

Cyclones

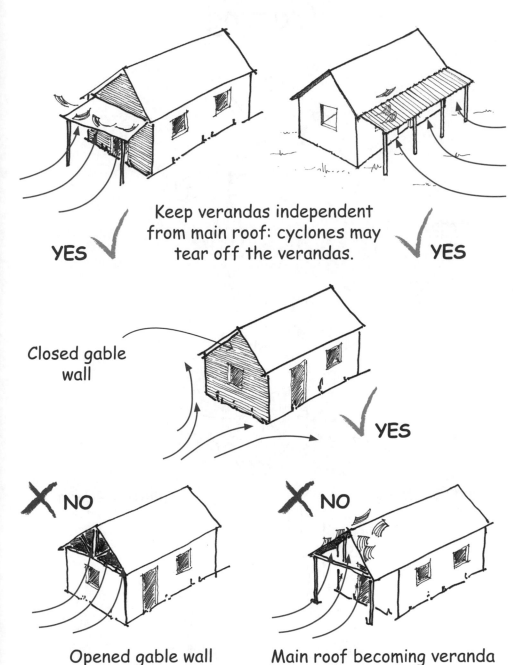

YES ✓ Keep verandas independent from main roof: cyclones may tear off the verandas. ✓ **YES**

Closed gable wall ✓ **YES**

✗ **NO** Opened gable wall

✗ **NO** Main roof becoming veranda

If a veranda is part of the main roof, then a cyclone could tear off the whole roof.

Fastening of the veranda framing

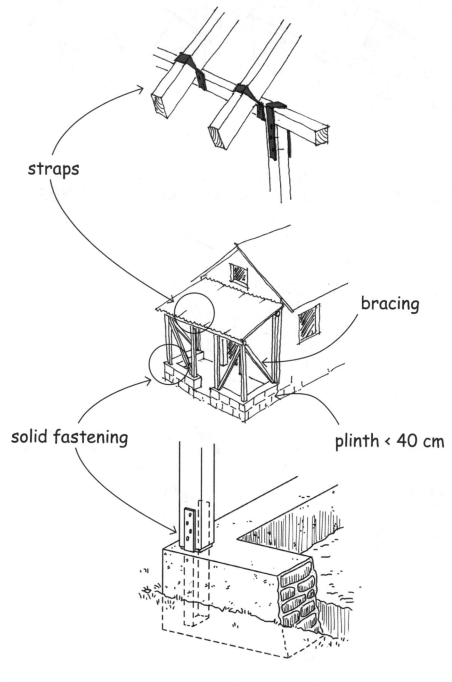

straps

bracing

solid fastening

plinth < 40 cm

Fastening of the roof structure

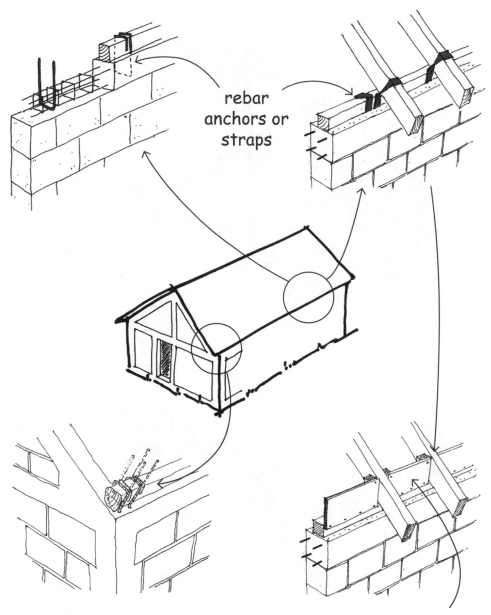

rebar anchors or straps

Solidly fasten the anchors or straps to the wood framing.

Close the spaces between trusses with a plank or a screen to deter insects.

Bracing

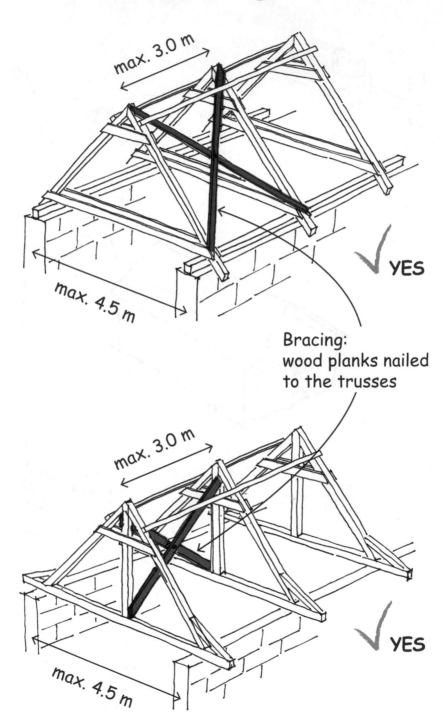

max. 3.0 m

max. 4.5 m

✓ YES

Bracing:
wood planks nailed
to the trusses

max. 3.0 m

max. 4.5 m

✓ YES

http://dx.doi.org/10.3362/9781780449883.014

14. RETAINING WALLS

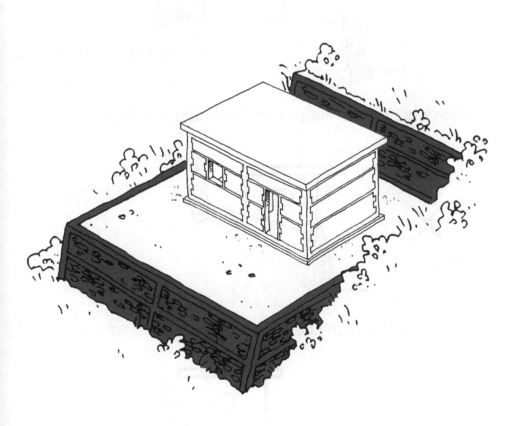

Where to build with retaining walls

A retaining wall is not a house wall. Water will and must leak through it.
A retaining wall is meant to hold back the ground.

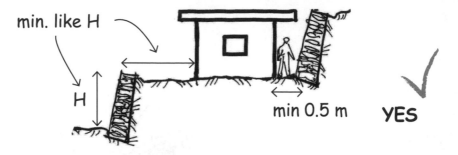

min. like H

H

min 0.5 m

YES

Don't build your house too close to a retaining wall.

NO

Don't build your house on top of a retaining wall.

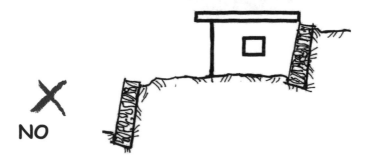

NO

Don't build your house against a retaining wall.

Rule 1 - Wall footing

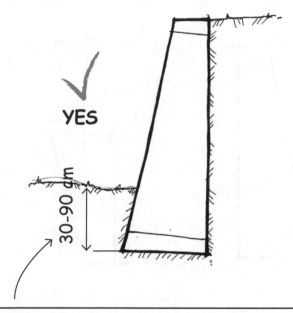

YES

30-90 cm

Height from bottom of wall to firm soil

- hard soil: 30 cm
- rammed soil: 30–60 cm
- soft soil: 60–90 cm

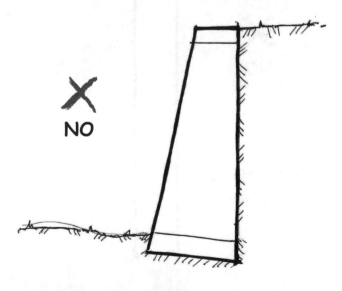

NO

Rule 2 - Slope of the wall (5:1)

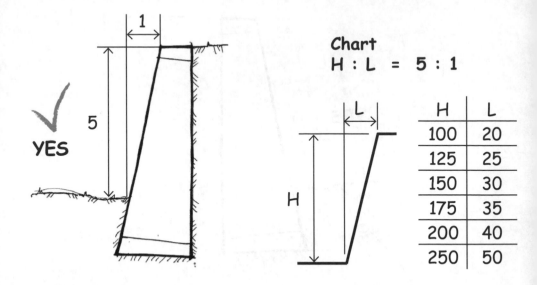

Chart
H : L = 5 : 1

H	L
100	20
125	25
150	30
175	35
200	40
250	50

Slope 1:5
Every time you go up 5 cm, move back 1 cm
Every time you go up 1 meter, move back 20 cm

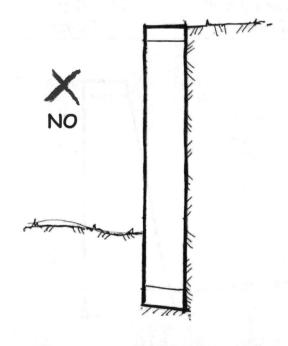

Rule 3 - Dimensions of the wall

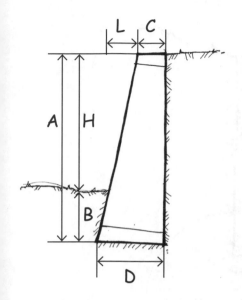

Height above ground (H):
H max = 2.50 m

Top (C): min 50cm
50 cm: H ≤ 150 cm
55 cm: H > 150 < 250 cm
60 cm: H ≥ 250 cm

Total height (A):
A = H + B
(-> B = 30-80 cm)

Wall base width (D) calculation:
The base of the wall (D) equals the total height (A) divided by 5, plus the top's width (C):

$$D = A/5 + C$$

Table

H	C	B	A	D
100	50	30-80	130-180	75-85
125	50	30-80	155-205	80-90
150	50	30-80	180-230	85-95
175	55	30-80	205-255	95-100
200	55	30-80	230-280	100-110
250	60	30-80	280-330	115-125

Rule 4 - Placing the stones

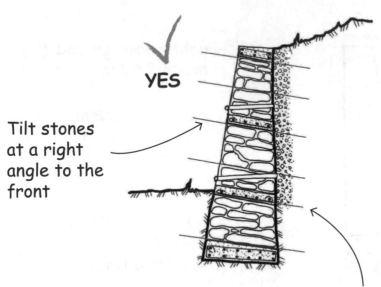

YES

Tilt stones at a right angle to the front

Place the stones on their flat faces and tilt them towards the back.

Place the stones at right angles to the wall's external face.

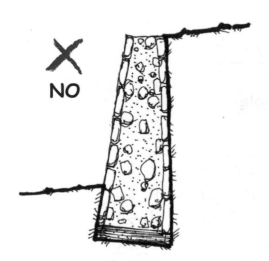

NO

Don't place the stones vertically.

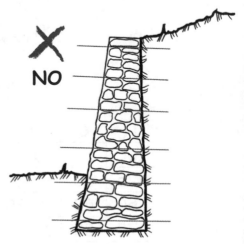

NO

Don't place the stones horizontally.

Rule 5 - Through-stones (or bands)

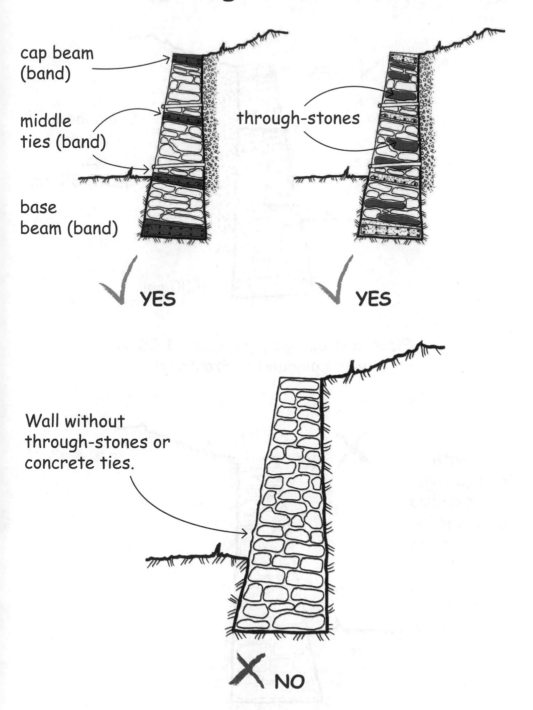

cap beam (band)

middle ties (band)

through-stones

base beam (band)

✓ YES   ✓ YES

Wall without through-stones or concrete ties.

✗ NO

Rule 6 - Drainage

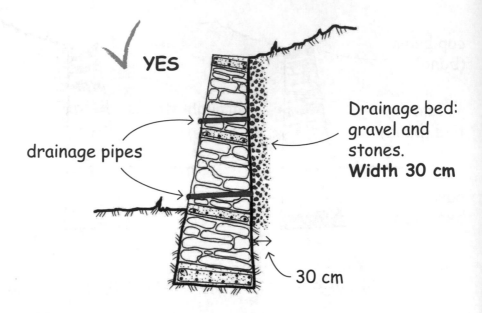

✓ YES

drainage pipes

Drainage bed:
gravel and
stones.
Width 30 cm

30 cm

Place a drainage pipe every 1.50 m
(vertically and horizontally)

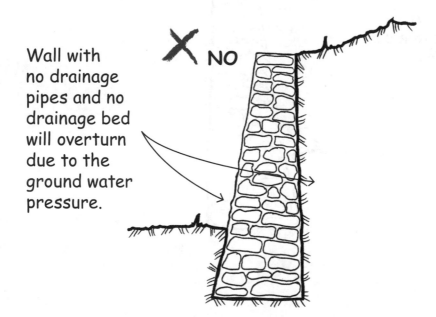

✗ NO

Wall with
no drainage
pipes and no
drainage bed
will overturn
due to the
ground water
pressure.

Retaining wall - Confining elements

These recommendations are for building a house on retaining walls: **Do it only if there is no other option.**

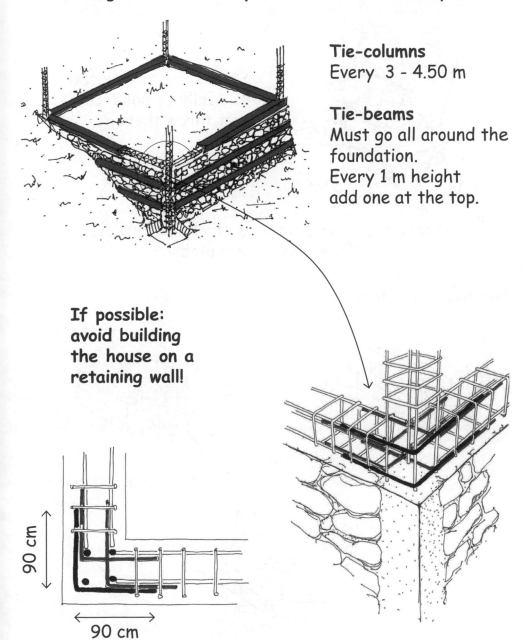

Tie-columns
Every 3 - 4.50 m

Tie-beams
Must go all around the foundation.
Every 1 m height add one at the top.

If possible: avoid building the house on a retaining wall!

90 cm

90 cm

119

Gabion retaining walls 1

Gabion walls consist of baskets woven with galvanised wire and carefully filled with stones.

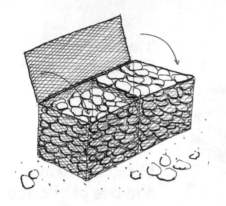

Stones must be placed carefully by hand. Don't just throw them in.

There are several ways to stack the baskets.
All are equally acceptable.

Method 1: in steps

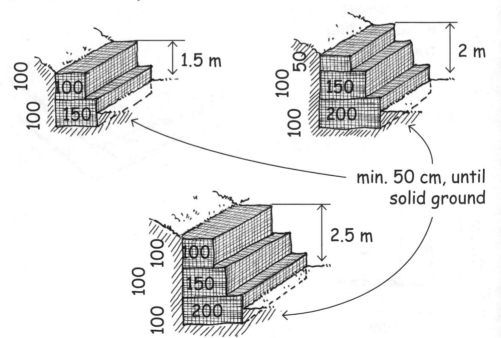

min. 50 cm, until solid ground

Gabion retaining walls 2

Method 2:
with a vertical face

Method 3:
with an inclined face

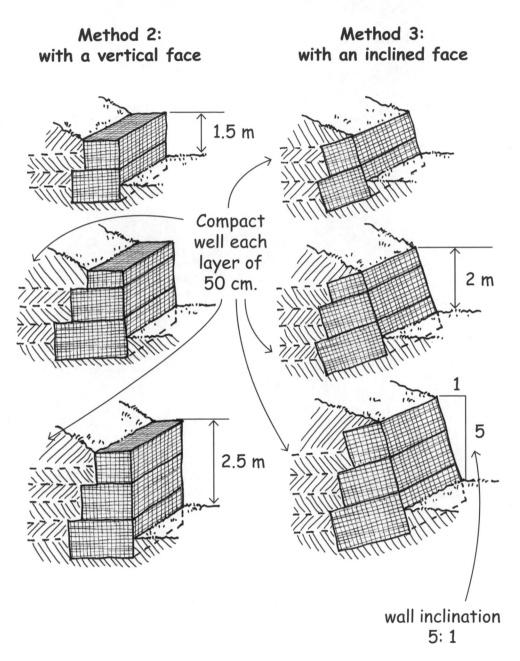

1.5 m

Compact
well each
layer of
50 cm.

2 m

2.5 m

1

5

wall inclination
5: 1

15. CONSTRUCTION DRAWINGS

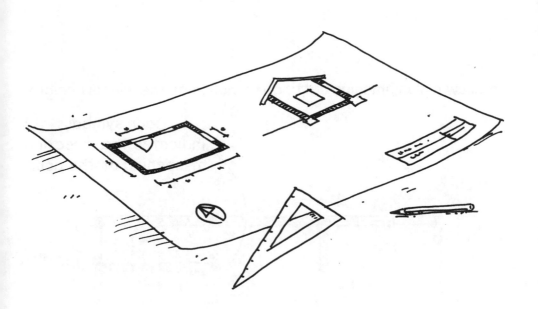

Reading plans

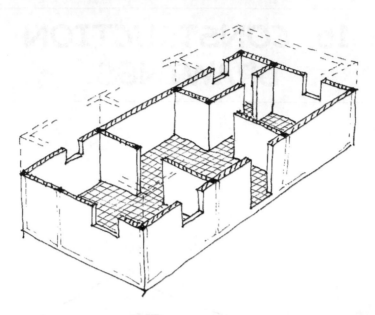

To draw a plan, imagine cutting the house at the window height.

Door symbol: indicates the direction of opening of the door.

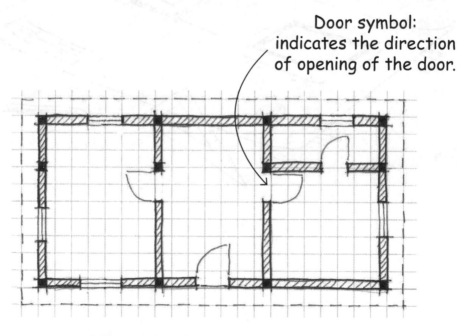

House plan (seen from the top).

Reading sections

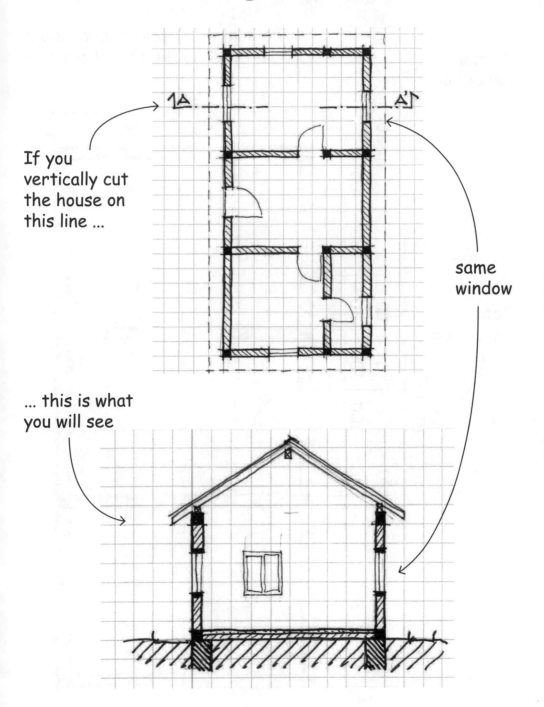

If you
vertically cut
the house on
this line ...

same
window

... this is what
you will see

Plan dimensions

The sum of all partial dimensions must result in the total dimension.

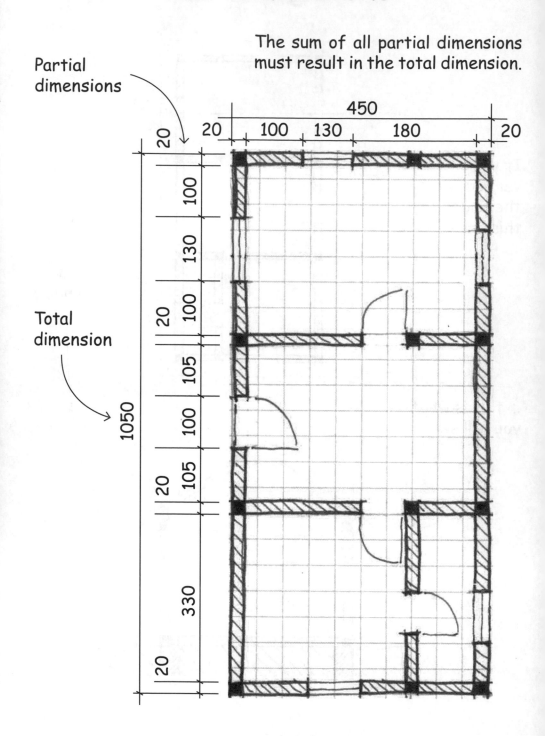

Partial dimensions

Total dimension

Section dimensions

Partial
dimensions

Partial
dimensions

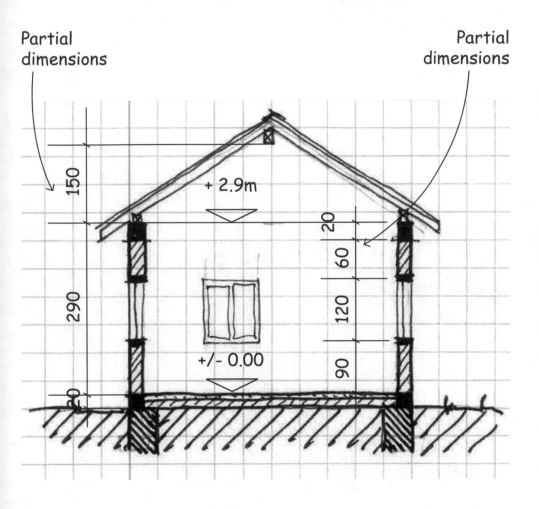